EXERCISES IN CERTIFICATE MATHEMATICS

1

G M Hocking B.Sc.
Head of Mathematics, Kingsfield School, Bristol
Assistant Examiner (O Level), London Board
Chief Examiner (CSE), South Western Examination Board

MACMILLAN

First published 1974

Published by
MACMILLAN EDUCATION LIMITED
London and Basingstoke

Associated companies and representatives throughout the world

Printed in Great Britain by
A. WHEATON AND CO.
Exeter

Preface

This book – and its companion 'Exercises in Certificate Mathematics 2' – is for C.S.E. and O Level students. It is planned for use in the final two years of their certificate course and intended to enrich and extend their mathematical experience and to develop confidence in the application of mathematical techniques and concepts, through challenge and success.

The questions in the two books cover almost all the requirements of certificate examinations in mathematics. They have been used in pilot schemes with C.S.E. and O Level candidates with abilities varying from somewhat below to substantially above average and their content and form modified in the light of this experience.

For the convenience of the user, the questions have been classified under topic headings. Answers are included at the end of the book, in the belief that candidates of this age can use answers sensibly.

The contents are designed to be both instructive and imaginative. The drudgery of over-repetitive questions has been avoided and the emphasis placed on providing a variety of question types and topics, that simultaneously stimulate and educate, in a lively and enlightening style.

It is not intended that any one student should attempt every question or indeed every exercise. For revision purposes, it is sufficient to try every other question in a relevant exercise, or every third question, or even every prime numbered question, filling in with further examples where the student is in difficulty or on unfamiliar ground. Such a procedure will provide a sufficient cross-section of experience in a particular topic. Students should be encouraged to make up questions for themselves, using the given examples as prototypes.

As ever, the instruction and guidance of a stimulating and understanding teacher is vital. For such a teacher the book provides, in compact form, a wealth of questions which he may select and readily augment to meet the needs of his particular pupils.

My thanks are due to staff and students who with zest and care have checked answers and suggested improvements in wording. Finally, I must record my gratitude to Mr Tim Pridgeon for his quiet faith and gentle guidance throughout its preparation.

G M HOCKING

Contents

6

TWO-DIMENSIONAL FIGURES

THREE-DIMENSIONAL FIGURES

Sets

Exercise 1

1 List the elements of (a) {days of the week}, (b) {months of the year with 31 days}, (c) {months of the year with 33 days}, (d) {prime numbers between 18 and 28}, (e) {multiples of 6}, (f) {factors of 30}.

Which one of the above is an infinite set? Which is a null set?

2 Write in an alternative form (a) {a e i o u}, (b) {red, green, amber}, (c) {v w x y z}, (d) {23, 46, 69, 92, ...}, (e) {triangle, quadrilateral, pentagon, ...}.

Which of the above are infinite sets?

3 Say whether the following are finite, infinite or null sets. (a) {Tuesdays in February},   (b) {numbers divisible by 15}, (c) {half-days in a year},   (d) {even prime numbers greater than 2},   (e) {square roots of 36},   (f) {multiples of 5}, (g) {odd numbers that factorise}.

4 R = {A B D F G}, S = {B C D E}, $\mathscr{E}$ = {letters of the alphabet}.

List R ∪ S, R ∩ S, R ∩ S', R' ∩ S.

What relation can you find between the four answers?

5 $\mathscr{E}$ = {days of the week}, M = {Monday, Tuesday, Wednesday}, T = {Tuesday, Wednesday, Thursday}.

List the sets M', M ∪ T, M ∩ T, M' ∪ T', (M ∩ T)'.

What relation is there between the final pair of answers?

6 T = {multiples of 3}, F = {multiples of 5}, S = {perfect squares}, $\mathscr{E}$ = {natural numbers}.

Give the three smallest elements in (a) each of the above sets, (b) T ∩ F, (c) T ∪ F, (d) T ∪ S, (e) T ∩ S, (f) T ∩ F ∩ S.

7 X and Y are the sets of points shaded in Fig. 1.1. Draw illustrations of X ∪ Y, X ∩ Y, (X ∩ Y)', X ∩ Y', X' ∪ Y.

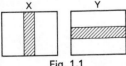

Fig. 1.1

Fig. 1.2

8 K, L and M are the shaded regions in Fig. 1.2.

(a) Illustrate K ∪ L, K ∩ L, L ∩ M, K ∪ L ∪ M.

8

(b) Take the circle as the universal set and illustrate: M′,
K′ ∩ L′,   (K ∪ L)′,   K′ ∪ L,   (K ∩ L′)′.

9   Say which of the following statements are true, and correct
those that are false.
   (a) {vowels} ⊂ {letters of the alphabet},   (b) {molars} ⊂
{teeth},   (c) {cards in a pack} ⊂ {hearts, clubs},   (d) {plain,
purl} ⊃ {stitches},   (e) {wickets, bails, gloves} ⊂ {cricket
gear}.

10   Correct the errors in the following statements.
   (a) {whole numbers} ⊂ {integers},   (b) {quadrilaterals} ⊃
{polygons},   (c) {isosceles triangles} ⊂ {equilateral triangles},
(d) {counting numbers} ⊂ {rational numbers},   (e) {cubes} ⊂
{cuboids},   (f) {rhombi} ⊂ {squares}.

11   R = {win, lose, draw}. How many subsets of R are there
with   (a) 1 element,   (b) 2 elements?
   How many subsets are there with at least one element?

12   P = {ace, king, queen, jack}. How many subsets of P are
there with   (a) 1 element,   (b) 2 elements,   (c) 3 elements?
   How many subsets of P are there other than P and ϕ?

13   List the subsets of   (a) {x, y},   (b) {r, s, t}.   How many
are there in each case?

14   How many subsets are there for   (a) {p, q, r, s},   (b) {v,
w, x, y, z}?

15   The number of subsets of a set with n elements is 2^n.
Calculate the number of subsets when there are 6, 7, 8 and 10
elements.
   How many subsets are there for the null set?

16   T = {John, Jill, Alison, Andrew, James, Jacqueline}. List
the subsets which contain: (a) girls only,   (b) a boy and a girl
only,   (c) teenagers with an initial letter A.
   How many subsets are there containing   (a) two boys and
two girls,   (b) two boys and one girl,   (c) teenagers with initial
letter J?

17   In Fig. 1.3 the figures show the number of elements in each
region. Use this information to find: $n(P)$, $n(Q)$, $n(P \cap Q)$,
$n(P \cup Q)$, $n(P')$, $n(P \cup Q')$.

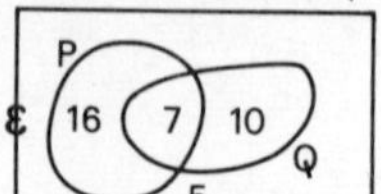

Fig. 1.3

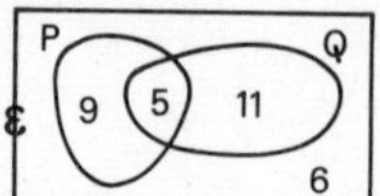

Fig. 1.4

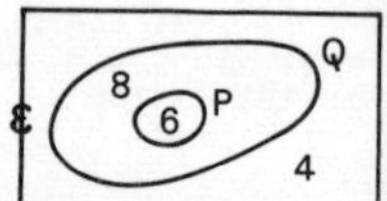

Fig. 1.5

18   Repeat question 17 for Figs. 1.4 and 1.5.

19 Use the information of Fig. 1.6 to find: $n(X)$, $n(Z)$, $n(X \cap Z)$, $n(\mathscr{E})$, $n(Y)$, $n(X \cap Y')$, $n(Z')$.

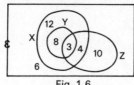

Fig. 1.6

20 $\mathscr{E} = \{\text{picture cards, including aces}\}$, $K = \{\text{kings}\}$, $S = \{\text{spades}\}$.
Find $n(\mathscr{E})$, $n(K)$, $n(S)$, $n(K \cap S)$.

21 $\mathscr{E} = \{\text{dominoes}\}$, $B = \{\text{dominoes with at least one blank section}\}$, $F = \{\text{dominoes with at least one 4-dot section}\}$.
Find $n(B)$, $n(F)$, $n(B \cap F)$, $n(B \cup F)$, $n(\mathscr{E})$, $n(B')$.

22 $n(P) = 15$, $n(Q) = 20$, $n(P \cup Q) = 30$ and $n(\mathscr{E}) = 45$.
Find $n(P \cap Q)$, $n(P')$, $n(P \cup Q)'$, $n(P \cap Q')$, $n(P \cup Q')$.

23 Draw Venn diagrams to illustrate sets S and T and the universal set and shade (a) $S \cup T$, (b) $S \cap T$, (c) T', (d) $S' \cap T$, (e) $T \cup S'$.

24 Make four copies of the Venn diagram in Fig. 1.7 and on the first three shade (a) $P \cap Q$, (b) $P \cup Q$, (c) $P \cap Q'$.

Fig. 1.7

On the fourth diagram add a set R such the $R \subset P$ and $R \cap Q = \phi$.

25 Draw Venn diagrams to illustrate the relation between (a) squares and rectangles, (b) circles and ellipses, (c) football players and goalkeepers, (d) prime numbers and whole numbers, (e) odd numbers and prime numbers.

26 Illustrate by Venn diagrams the relations between (a) triangles, isosceles triangles and right angled triangles, (b) prisms, cylinders and cones, (c) cones, pyramids and spheres, (d) quadrilaterals, squares and kites, (e) parallelograms, rectangles and rhombi.

27 Use Venn diagrams to illustrate the following relations. (a) $A \subset B$, (b) $A \cap B = \phi$, (c) $A \cup B = B$, (d) $(A \cap B)' = A'$.

28 L = {factors of 28},   M = {factors of 24}, $\mathscr{E}$ = {whole numbers}. Draw a Venn diagram for L, M and $\mathscr{E}$. Place the following numbers in their appropriate positions on your diagram: 6,   7,   4,   2,   5,   3.

29 P = {1, 3, 5, 7},   Q = {5, 10, 15, 20},   R = {5, 1, 7, 3}, S = {10, 20}.
Say whether the following statements are true or false.
(a) P = R,   (b) n(P) = n(Q),   (c) S $\subset$ Q,   (d) P $\cap$ Q = ϕ, (e) 5$\in$(P $\cap$ Q),   (f) P $\cup$ R = P,   (g) Q $\cup$ S = S.

30 $\mathscr{E}$ = {3, 4, 5, 6 . . . . .},   E = {even numbers},   O = {odd numbers},   S = {multiples of 6},   T = {multiples of 3}.
(a) State the relation between E and O and between T and S. (b) List the four smallest elements in E $\cap$ T.   (c) State a relation between E, T and S.   (d) Draw a Venn diagram to illustrate the five sets.   (e) Which of the following are null sets? E $\cap$ O,   E $\cup$ O,   S $\cap$ O,   T $\cap$ E.

31 $\mathscr{E}$ = {bicycles},   T = {tandems},   V = {veteran cycles}.
(a) Illustrate the above sets by a Venn diagram.   (b) What is represented by T', V', T' $\cap$ V'?   (c) Add to your diagram, B = {British built bicycles}.

32 58 business men leapt from the train. 40 carried newspapers, 30 carried umbrellas and 20 carried both. How many carried neither?

33 On Sunday, 163 people attended morning service and 143 people attended evening service. 260 different people attended the services in all. How many went to two services that Sunday?

34 Out of 30 eighteen-year-olds 20 had driven a car, 17 had driven a motor cycle and 4 had not driven either. How many had driven both vehicles?
A year later 22 had driven cars, 21 motor cycles and 14 both. How many non-drivers were there left?

35 Of the 40 members of a Winter Sports Club twice as many skate as ski. Half of those who ski also skate. 5 members are new and have yet to take part in either activity. How many members ski?

36 Out of 30 different numbers, 12 are divisible by 3, 14 are divisible by 5, and 6 are divisible by 15. How many numbers are there that are divisible neither by 3 nor by 5?

37 There are 48 families in a village. 31 own a cat and 26 own a dog. Twice as many families own both pets as those that have neither.
(a) Call the number of families with no pets 'n' and express

the number with both pets in terms of *n*.   (b) Find expressions for the numbers with dogs only and cats only.   (c) Form an equation in '*n*' and solve it.   (d) How many families have only a cat?

38   The 'Hotel Supreme' opens every month of the year except November, January and February. The 'Hotel Tip Top' opens from April to November inclusive.
   Form sets S and T to show the months when the hotels are open. List, and explain the meaning of,   S ∩ T,   S ∪ T,   T′, S ∩ T′.

39   The 'Crusty Bread Co' delivers on Mondays, Tuesdays and Fridays. 'Better Loaf Ltd.' delivers on Tuesdays, Wednesdays, Fridays and Saturdays.
   Form sets C and B to show the delivery days of the two companies. Explain the information given by   C ∪ B,   C ∩ B, B′,   (C ∪ B)′,   C′ ∩ B.

40   Explain why the information $n(A ∪ B) = 60$, $n(A) = 40$, $n(B) = 70$ and $n(A ∩ B) = 12$ is impossible.
   Assume there is only one error and suggest two possible corrections.

Whole Numbers

Exercise 2

1   Write in short form   (a) eighty-eight thousand five hundred and two,   (b) three thousand and fifty-three,   (c) fourteen million and six,   (d) four-hundred-and-four thousand six hundred and five.

2   Find the difference in value between   (a) the 5's in 585, (b) the 6's in 26 361,   (c) the 7's in 7 117,   (d) the 8's in 83 828, (e) the 9's in 999.

3   Find the set of numbers formed by adding together any two elements of {135, 226, 384}. What is the relation between the average of the numbers in the original set and the average of the numbers in the new set?

4   Find the set of numbers formed by finding the difference between any two elements of {111, 234, 356}.

5   List the numbers formed by adding together the six pairs of elements of   {12, 14, 18, 24}.
   Repeat for   {25, 46, 58, 91)   and   {68, 135, 226, 381}.

6   Compare the averages of the original four numbers in each of the sets in question 5 with the averages of the final six numbers. What do you notice? Can you explain this relation?

7   List S = {multiples of 6 between 31 and 101} and
      E = {multiples of 8 between 31 and 101}.
   Find S ∩ E, and describe this set in words.

8   Make a table of the prime numbers p and q such that
(a) $p + q = 24$,   (b) $p + q = 40$,   (c) $p + q = 43$,
(d) $p + q = 55$ and   (e) $p + q = 80$.

9   Find the first three pairs of prime numbers p and q such that
$p - q$ equals   (a) 8,   (b) 12,   (c) 26,   (d) 34.
   What difficulties do you meet if $p - q = 7$?

10   List the whole numbers w and v such that   (a) $w + v = 20$,
(b) $wv = 24$,   (c) $w^2v = 144$.
   What difficulty is there in listing the whole numbers if
$\dfrac{w}{v} = 24$? Give just four of the possible pairs.

11   Make a table of multiples between 159 and 199 of   (a) 6,
(b) 7   and   (c) 8. Use your answers to find multiples of 42, 24 and 56, in this range.

12   Find the elements of
      S = {perfect squares from 1 to 101} and
      C = {perfect cubes from   1 to 101}
   Are there any elements in   S ∩ C?

13   What are the smallest and largest numbers that can be formed from the figures   (a) 5 6 7,   (b) 8 2 4   and   (c) 3 3 5? In each case find the difference between the largest and the smallest numbers. Check that these differences are multiples of 99. Why is this?

14   Find the smallest number that must be added to 8 272 to make it divisible by 53. What are the next two smallest numbers?

15   O_1 and O_2 are odd numbers. E_1 and E_2 are even numbers. Say whether the following are odd or even: (a) $O_1 + E_1$,
(b) $E_1 + E_2$,   (c) $2 \times O_1$,   (d) $5 \times O_2$,   (e) $O_1{}^2$,
(f) $O_1{}^2 + O_2{}^2$,   (g) $O_1 + O_2$,   (h) $O_1 \times O_2$.

16   Write down the squares of the whole numbers from 11 to 15 inclusive. Find   (a) their sum,   (b) their average,   (c) the differences between successive squares. Use this final result to help you write down the squares of the next five whole numbers.

17 Test whether the following numbers are divisible by 3, 4, 5 or 7: 120, 285, 420, 630, 1 080, 1 155.
Use your answers to tell whether they are divisible by 15, 28, 35 or 60.

18 Express the following as the product of prime numbers and hence find their HCF and LCM: (a) 24, 28; (b) 36, 54; (c) 48, 84; (d) 52, 182; (e) 45, 60, 150; (f) 44, 66, 242.

19 What is the first prime number after (a) 10, (b) 100, (c) 1 000?

20 Find the smallest whole number that 2 205 must be multiplied by to make it (a) a perfect square, (b) a perfect cube, (c) a multiple of 360.

21 Three lighthouses flash their lights at intervals of 54, 60 and 72 seconds. If they flash together at 20 00 h when will they next flash together?

22 Repeat question 21 for intervals of (a) 48, 60 and 72 seconds, (b) 36, 60 and 84 seconds, (c) 42, 56 and 70 seconds.

23 (a) The average of four numbers is 5. What number must be added to increase the average to 7?
(b) The average of fifty-seven numbers is 85. What number must be added to increase this to 87?

24 The average of seventeen numbers is 43. What number must be added to (a) increase this average to 45, (b) decrease the average to 40?

25 The attendances at three successive home games of a football club were 18 327, 21 293 and 20 755. Find (a) the total attendance, (b) the average attendance, (c) the attendance at the next game if this brought the average to 21 000.

26 John Rashley is always making mistakes. What is his error if he (a) adds 352 instead of 325, (b) subtracts 46 instead of 64, (c) adds 831 instead of 183, (d) multiplies 145 by 32 instead of 23?

27 . . . . (a) Draw the next three patterns in this
         .   . .   series
            .

and count the number of dots in each of the six terms. (b) Use your results to predict the number of dots in the 7th, 8th and 12th terms. (c) Draw a diagram to show why two successive triangular numbers form a square number.

28   Copy and complete the following calculations.

(a) $38* +$ (b) $4*7 -$ (c) $*23 \times$ (d) $8*)\overline{290}^{\,3.*}$
    $2*5$ $*18$ $8*$ $2*9$
   $\overline{*47}$ $\overline{17*}$ $\overline{3*840}$ $\overline{**5}$
                                              $4*3$ $41*$
                                             $\overline{*****}$ $\overline{000}$

29   Find the missing terms in the series:

(a) 5, 9, 13, —, —, 25, 29   (b) 1, 4, 9, —, 25, —,
(c) 2, 5, 10, 17, —, —,   (d) 6, 18, 54, —, 486, —,
(e) 1, 5, 14, 30, —, —,   (f) 2, 2, 4, 6, 10, 16, —, —.

Fractions

Exercise 3

1   Using squared paper draw 4×3 rectangles and shade squares to illustrate $\frac{1}{3}, \frac{1}{4}, \frac{1}{2}$ and $\frac{5}{6}$.

2   Use 6×4 rectangles and shade illustrations of $\frac{1}{6}, \frac{1}{3}, \frac{1}{8}, \frac{3}{8}, \frac{5}{12}$ and $\frac{11}{12}$. What fractions do the unshaded parts represent?

3   Shade squares in 5×4 rectangles to illustrate $\frac{1}{2}, \frac{3}{4}, \frac{3}{10}, \frac{1}{5}, \frac{3}{5}, \frac{13}{20}$ and $\frac{17}{20}$.

4   Draw pairs of 4×3 rectangles to illustrate $\frac{3}{12} = \frac{1}{4}$; $\frac{2}{6} = \frac{1}{3}$; $\frac{2}{3} + \frac{1}{12} = \frac{3}{4}$; $\frac{1}{3} - \frac{1}{4} = \frac{1}{12}$; $\frac{1}{3}$ of $\frac{1}{2} = \frac{1}{6}$; $\frac{1}{2} \div 3 = \frac{1}{6}$; $\frac{1}{3} \div \frac{1}{6} = 2$.

5   Use pairs of 6×3 rectangles to illustrate $\frac{2}{9} = \frac{4}{18}$; $\frac{1}{2} = \frac{3}{6}$; $\frac{1}{6} + \frac{1}{3} = \frac{1}{2}$; $\frac{1}{2} - \frac{5}{18} = \frac{2}{9}$; $\frac{1}{3}$ of $\frac{1}{6} = \frac{1}{18}$; $\frac{1}{3} \div 3 = \frac{1}{9}$; $\frac{1}{6} \div \frac{1}{18} = 3$.

6   Copy and complete the following:

(a) $\frac{2}{3} = \frac{}{15} = \frac{32}{} = \frac{}{75} = \frac{}{3000}$,   (b) $\frac{5}{8} = \frac{20}{} = \frac{}{56} = \frac{45}{} = \frac{}{400}$
(c) $\frac{3}{4} = \frac{18}{} = \frac{}{28} = \frac{27}{} = \frac{}{160}$,   (d) $\frac{7}{9} = \frac{}{45} = \frac{63}{} = \frac{}{90} = \frac{}{999}$
State three more fractions equivalent to each of the above.

7   State the simplest element in each of the following sets of equivalent fractions. Name three further elements in each set.

(a) $\{ \ldots \frac{15}{25}, \frac{30}{50}, \frac{36}{60}, \ldots \}$,   (b) $\{ \ldots \frac{18}{27}, \frac{38}{57}, \frac{120}{180}, \ldots \}$
(c) $\{ \ldots \frac{66}{72}, \frac{99}{108}, \frac{121}{132}, \ldots \}$,   (d) $\{ \ldots \frac{25}{45}, \frac{30}{54}, \frac{125}{225}, \ldots \}$
(e) $\{ \ldots \frac{9}{24}, \frac{15}{40}, \frac{33}{88} \ldots \}$,   (f) $\{ \ldots \frac{14}{24}, \frac{49}{84}, \frac{245}{420}, \ldots \}$.

8 State any three elements in the set of fractions equivalent to $\frac{1}{4}$, $\frac{1}{3}$, $\frac{2}{5}$, $\frac{3}{8}$, $\frac{5}{9}$, $\frac{4}{5}$, $\frac{7}{10}$, $\frac{11}{12}$.

9 Simplify: $\frac{1}{3} + \frac{1}{4}$; $\frac{5}{6} + \frac{7}{8}$; $2\frac{3}{10} + 4\frac{1}{4}$; $\frac{4}{5} - \frac{2}{3}$; $\frac{5}{12} - \frac{7}{18}$; $3\frac{4}{5} - 1\frac{1}{3}$; $3\frac{1}{3} - 1\frac{4}{5}$; $2\frac{1}{2} + 3\frac{1}{6} - 1\frac{1}{4}$; $1\frac{1}{3} - 2\frac{3}{8} + 3\frac{1}{2}$.

10 Calculate: $\frac{1}{3} \times \frac{1}{3}$; $\frac{1}{2} \times \frac{1}{4}$; $\frac{2}{3} \times \frac{3}{4}$; $\frac{4}{5} \times \frac{15}{16}$; $1\frac{3}{4} \times \frac{2}{7}$; $1\frac{1}{2} \times 1\frac{1}{6}$; $3\frac{1}{7} \times 4\frac{2}{3}$; $7\frac{1}{2} \times 1\frac{1}{3}$.

11 Evaluate: $1 \div \frac{1}{5}$; $6 \div \frac{1}{8}$; $4 \div \frac{2}{5}$; $\frac{1}{2} \div \frac{1}{4}$; $\frac{3}{5} \div \frac{9}{10}$; $4\frac{1}{6} \div \frac{5}{8}$; $3\frac{3}{4} \div 3\frac{1}{8}$; $2\frac{1}{4} \div 1\frac{7}{8}$.

12 Reduce to a single fraction: $3\frac{2}{3} \times 1\frac{1}{2} \div 1\frac{5}{6}$; $(4\frac{1}{5} \div 2\frac{1}{3}) \times 1\frac{2}{3}$; $\frac{1}{2} + \frac{1}{3} \div \frac{1}{6}$; $(\frac{1}{2} + \frac{1}{3}) \div \frac{1}{6}$; $\frac{1}{2} \div \frac{1}{3} + \frac{1}{6}$; $\frac{1}{2} \div (\frac{1}{3} + \frac{1}{6})$.

13 If $p = 2\frac{1}{3}$ and $q = 1\frac{1}{2}$, find the value of $p + q$; pq; $p \div q$; $(p - q)/pq$.

14 If $r = 3\frac{1}{4}$ and $s = 2\frac{3}{5}$, find the value of: $r - s$; rs; $r \div s$; $(r + s)/(r - s)$; r^2/s^2.

15 Express the following fractions of a pound (£) in pence (p): $\frac{1}{2}$, $\frac{1}{4}$, $\frac{1}{8}$, $\frac{3}{5}$, $\frac{7}{10}$, $\frac{9}{20}$, $\frac{16}{25}$, $\frac{31}{50}$, $\frac{49}{50}$.

16 Express in minutes the following fractions of an hour: $\frac{1}{5}$, $\frac{5}{6}$, $\frac{3}{4}$, $\frac{3}{10}$, $\frac{7}{12}$, $\frac{2}{3}$, $\frac{5}{8}$.

17 Express the following times as fractions of a week (7 days): 12 hours; 35 hours; 56 hours; the whole of Monday and two thirds of Tuesday; the weekend from 5 00 pm on Friday to 8 00 am on Monday.

18 If $\frac{5}{6}$ of the earth's surface is water and $\frac{2}{3}$ of the remainder is uninhabitable, what fraction is dry and habitable?

19 A VHF radio sells for £35. Tax accounts for $\frac{1}{5}$ of this and profit for $\frac{1}{4}$ of the remainder. What is the cost price of the radio?

20 Last year a man was able to save $\frac{1}{10}$ of his income. This year his expenditure has increased by $\frac{1}{4}$ and his income by $\frac{1}{5}$. What fraction of his new income can he now save?

21 ABCD is a rectangle and P, Q, R and S are the mid-points of its sides. Express as a fraction of the whole rectangle the area of the triangles ABD; ASP; PDC; PSR; DRQ.

22 At the Uckbridge School Concert, $\frac{1}{4}$ of the time was taken up by the choir, $\frac{1}{3}$ by the orchestra and $\frac{3}{10}$ by solo performers. The most popular item – the interval – accounted for the remainder. How long was the interval, if the proceedings lasted $1\frac{1}{2}$ hours? Illustrate the division by a pie chart.

23   Oxygen was fed into a space capsule at a rate that would fill it in 15 minutes. The astronauts used the oxygen at a rate that would empty it in 12 minutes. If the capsule started full, how long would it be before it was empty?

24   An escalator rises from the first floor to the second floor in 45 seconds. For a lark, a man runs down the escalator at a speed that would take him from the second floor to the first in 30 seconds, if the escalator were stopped. How long does it take him with the escalator moving?

Decimals and Percentages

Exercise 4

1   $2 \cdot 83 = 2 + \frac{8}{10} + \frac{3}{100}$. Write in full, $1 \cdot 32$, $2 \cdot 46$, $3 \cdot 052$, $6 \cdot 708$, $9 \cdot 4005$, $5 \cdot 0302$.

2   Write in short form: $8 + \frac{3}{10}$, $9 + \frac{7}{100}$, $7 + \frac{2}{10} + \frac{6}{100}$, $3 + \frac{7}{100} + \frac{5}{1000}$, $8 + \frac{2}{10} + \frac{3}{1000}$.

3   Write in decimal form: $2\frac{7}{10}$, $3\frac{5}{100}$, $4\frac{29}{100}$, $12\frac{81}{100}$, $9\frac{192}{1000}$, $14\frac{31}{1000}$, $10\frac{206}{1000}$.

4   Simplify: $3 \cdot 62 \times 10$, $29 \cdot 8 \times 100$, $3 \cdot 14 \times 200$, $0 \cdot 015 \times 30$, $15 \cdot 3 \div 100$, $1 \cdot 58 \div 1\,000$, $0 \cdot 837 \div 10$.

5   Calculate: $(0 \cdot 3)^2$, $(0 \cdot 8)^2$, $(0 \cdot 5)^2$, $(0 \cdot 02)^2$, $(0 \cdot 04)^3$, $\sqrt{0 \cdot 04}$, $\sqrt{0 \cdot 09}$, $\sqrt{0 \cdot 16}$, $\sqrt[3]{0 \cdot 000125}$.

6   In the table opposite, the totals of each row and each column are calculated. These are checked by finding the totals of the final row and final column. Carry out similar calculations for:

7·6	5·9	13·5
2·8	3·2	6·0
10·4	9·1	19·5

17·6	0·44
3·2	0·16

19·2	1·2
4·8	0·03

28·8	3·2
4·8	0·16

16·5	0·22
5·5	0·11

8·82	1·4
2·1	0·7

7   Repeat question 6 but replace addition by   (a) subtraction, (b) division.

8   $p = 16 \cdot 8$, $q = 0 \cdot 7$, $r = 2 \cdot 8$.   Find: $p + q + r$, pq, pr, qr, $p \div r$, $p \div q$, $r \div q$, r^2, q^3, p^2.

9 $r = 49{\cdot}2$, $s = 12{\cdot}3$, $t = 0{\cdot}12$. Find: $r - s$, $s - t$, $r - (s - t)$, $(r - s) - t$, rt, rs, $r \div t$, $r \div s$, $s \div t$, $t \div s$, t^2, s^2.

10 Write the numbers
$$34{\cdot}458, \quad 18{\cdot}317, \quad 5{\cdot}062, \quad 0{\cdot}0382$$
(a) correct to 2 decimal places,   (b) correct to 1 decimal place, (c) to 3 significant figures,   (d) to 1 significant figure.

11 Express as recurring decimals: $\frac{1}{3}$, $\frac{2}{3}$; $\frac{1}{9}$, $\frac{2}{9}$, $\frac{4}{9}$; $\frac{5}{11}$, $\frac{7}{11}$, $\frac{9}{11}$; $\frac{4}{7}$, $\frac{5}{7}$, $\frac{6}{7}$.

12 Write the answers to question 11,   (a) correct to 3 and (b) correct to 2 decimal places.

13 Consider the following:

Fraction	$F = 0{\cdot}36363\dot{6}\ldots$
Hence	$100F = 36{\cdot}363636\ldots$
Subtracting	$99F = 36$ Hence $F = \frac{36}{99} = \frac{4}{11}$.

Use this – or any other method – to find the value of: $0{\cdot}121\dot{2}\ldots$, $0{\cdot}31313\dot{1}\ldots$, $0{\cdot}5454\ldots$, $0{\cdot}1818\ldots$, $0{\cdot}666\ldots$, and $0{\cdot}432432\ldots$.

14 Write as a fraction and as a decimal fraction: 75%, 55%, 72%, $87\frac{1}{2}\%$, 48%, 85%, $2\frac{1}{2}\%$, 120%, 265%.

15 Express as a percentage and as a decimal fraction: $\frac{1}{2}$, $\frac{1}{4}$, $\frac{3}{4}$, $\frac{1}{8}$, $\frac{3}{8}$, $\frac{5}{8}$, $\frac{7}{8}$, $\frac{1}{10}$, $\frac{3}{10}$, $\frac{7}{10}$, $\frac{3}{20}$, $\frac{9}{20}$, $\frac{13}{20}$.

16 Write as a fraction and as a percentage: $0{\cdot}7$, $0{\cdot}64$, $0{\cdot}35$, $0{\cdot}375$, $0{\cdot}95$, $1{\cdot}45$, $2{\cdot}72$, $3{\cdot}15$.

17 A reduction of 30% was offered in a sale. Write 'sale price' tickets for goods originally costing: £24, £16, £12, £11, £9·50, £6·55, 96p, 54p.
    What was the original price of a hat with sale price 49p?

18 Sensational reductions in LP records are advertised. If the reduction is 48% how much is paid for records originally priced at: £1·80, £1·50, £1·20, 80p?
    What was the original price of records sold for (a) 39p, (b) £1·04?

19 The Treadwell Tyre Company offers a 30% discount on new tyres and a 35% discount on remoulds. How much does a customer pay for (a) 2 new tyres at £5 each,   (b) 2 remoulds at £4·50 each,   (c) 3 remoulds at £3·60,   (d) 4 new tyres at £5·30?
    On the average, the lifetime of a remould is 75% of that of a new tyre. Which is the better buy, new at £5·40 or remould at £4·60?

20 Gryptite Adhesives Ltd decide to pay £2 per week extra to their employees. What percentage increase is this to a worker earning: £10, £15, £20, £25, £36 and £48 per week?

Decimal Currency— Commercial Applications

Exercise 5

Note: Many of these questions can be worked using hand calculators, if these are available.

1 Take the cost of gas as 22p per therm and make out bills for 15, 18, 24, 28 and 36 therms.

2 The 'Silver Star Gas Tariff' charges a basic £3·25 per quarter plus 9·5p per therm. Which bills in question 1 would be cheaper on the Silver Star scheme?

3 Make out bills – using the Silver Star Tariff – for householders who used 48, 58, 66, 78 and 128 therms in a quarter.

4 Are any of the bills in question 3 cheaper on the 'Gold Star Tariff'? This charges a basic £6·50 per quarter plus 7·5p per therm.

5 Would you recommend the Silver Star or Gold Star schemes to a family using an average of 200 therms per quarter?

6 Calculate quarterly bills – on the Gold Star Tariff – for John Household who used 230 therms in the Spring, 52 in the Summer, 150 in Autumn and 360 therms in the Winter quarters.

7 Electricity is charged as follows:
First 60 *units* 4·5p each. *Remaining units* 0·8p each.
Complete bills for 42, 142, 400, 950 and 1 025 units.

8 Telephone charges are £4 per quarter for a shared line and £5 for a separate line. 1p per dialled unit on STD and 1·5p per unit if not on STD. Complete bills for the following.

Line shared	Yes	Yes	No	No	Yes	No
STD	Yes	No	Yes	No	No	Yes
No of units	320	260	442	296	664	787.

9   Calculate the separate and total costs of the following items of club equipment.
   16 coffee tables @ £2·25; 42 chairs @ £3·60; 85 coat hooks @ 7½p; 120 table tennis balls @ 3½p.

10   Find the discount on the total cost in question 9, at (a) 10%,   (b) 15%,   (c) 12½%.

11   Find the cash price of the following washing machines if there is a discount of 20%.   The machines' marked prices are £60,   £75,   £80,   £106   and   £120.

12   Repeat question 11 for discounts of   (a) 15%,   (b) 12½%, answers to the nearest 5p.

13   The Wreckam Rovers Supporters Club arrange a coach trip to an away match. The cost is £28·80. What should each supporter pay if   (a) 40,   (b) 36 travel?
   How many travel if the price is 90p?

14   The admission charge to Beargarden Zoo is 24p for adults, half price for children. Find the charge for   (a) father, mother and 3 children,   (b) father, mother, aunty, uncle and 7 children,   (c) a coach load of 26 cub scouts and 2 scout leaders.

15   Find the saving on the charges in question 14 if there is a reduction of 20% for parties of 8 and over.

16   The price of admission to a castle is 12p with 4p extra for the jewel room and 6p for the dungeons. Children are half price.
   Find the cost for   (a) a party of 15 seeing the castle only, (b) father, mother and 2 children, if mother went to the jewel room and father took the children to the dungeons,   (c) a party of 28 who went everywhere,   (d) a party of 30 who went everywhere, except to the dungeons.

17   The excursion fare to Slapton-on-Sea is 42p for adults and half that price for children. There is a reduction of 20% for parties of 8 and over. Find the cost for   (a) 2 adults and 1 child, (b) 4 adults and 3 children,   (c) 4 adults and 6 children,   (d) 8 adults and 18 children.

18   Under the conditions of question 17, what is the maximum number of people that can travel for   (a) £2 and   (b) £4, if there must be at least one adult to every 5 children or less?

19   Find the Simple Interest on   (a) £700 for 3 years at 5%, (b) £340 for 4 years at 8%,   (c) £250 for 6 years at 7½%, (d) £920 for 5 years at 6¼%,   (e) £630 for 6 months at 18%.

20   Find the rate at which   (a) £240 earns £30 interest in 5 years,   (b) £360 becomes £420 in 4 years.

21 Find the time in which (a) £300 amounts to £360 at 4%, (b) £150 amounts to £180 at $2\frac{1}{2}$%.

22 Find the principal that earns (a) £60 in 5 years at 4%, (b) £42 in 4 years at $3\frac{1}{2}$%.

23 Calculate the Compound Interest on (a) £400 for 3 years at 4%, (b) £200 for 3 years at 6%, (c) £600 for 2 years at 4% followed by 1 year at 5%, (d) £540 for 1 year at 6% and 2 years at 5%.

24 A £4 000 house is estimated to increase in value by 6% each year. Find its value after 3 years – to the nearest £50.

25 Repeat question 24 for a £5 000 house appreciating at 5% and a £30 000 house appreciating at 2% over 3 years.

26 The number of weeds in a field increases by 20% each week. Starting with 200 weeds, how many are there in 4 weeks time? Answer to the nearest 10.

27 Cars are estimated to decrease in value by 20% in the first two years and 15% in the next year. Find the value after three years of cars costing £800, £1 200, £1 500 and £1 850. Answers to the nearest £10.

28 A caravan depreciates in value by 20% in the first year, 15% in the second and 10% in subsequent years. Find the value after 4 years of caravans costing £400, £500, £750 and £1 250. Answers to the nearest £10.

29 A man borrows £800 at 4% compound interest and repays £150 at the end of each year. Find the amount owing after his third repayment. Answer to the nearest penny.

30 Repeat question 29 for (a) £200 at 5% repaying £60 (b) £2 000 at 6% repaying £300 each year.

31 Complete the profit and loss table below.

Cost Price	£25	£80	£35	£48		
Selling price	£28	£90			£54	£66
Profit %			20%	15%	20%	10%.

32 A record player costing £32 is sold for £36. Calculate the profit per cent. At what price must it be sold to give a 15% profit?

33 A carpenter buys lengths of wood at 5p each, machines them at 25p for four and sells them as rolling pins at £1·35 for ten.

Find   (a) the cost of 20 pins,   (b) the cash profit on 20 pins,
(c) his percentage profit.

34   A canoe was sold for £23 at a profit of 15%. What would
have been the profit if it had been sold at £25?

Social Applications of Arithmetic

Exercise 6

INSURANCE

1   Find the cost of insuring, at 60p per £100,   (a) a tape
recorder value £40,   (b) an antique vase value £65,   (c) lug-
gage of value £120,   (d) hi-fi equipment value £190,   (e) jewellery of value £1 250.

2   Take the premium for House Insurance as $2\frac{1}{2}$p for £10 and
that for contents as 5p for £10. Hence find the cost of insuring
the following houses and their contents.

| *House value* | £2 000 | £3 500 | £4 400 | £6 800 | £8 000 |
| *Contents value* | £ 800 | £1 200 | £1 800 | £2 200 | £3 200. |

3   Exhibition stands at Expo 75 were insured at a premium of
$7\frac{1}{2}$p in the £. Find the insurance costs for stands of value
£2 000,   £1 800,   £2 600,   £4 200   and   £5 500.

4   Find the cost of Rain Insurance at $12\frac{1}{2}$p in the £ for   (a) a
flower show with £340 cover,   (b) a tennis tournament with
£50 cover,   (c) a garden fete with £120 cover,   (d) an agri-
cultural show with £4 800 cover.

WAGES AND SALARIES

5   Calculate the monthly payment for annual salaries of
£480,   £840,   £660,   £1 260,   £1 500,   £1 680   and   £1 884.

6   Calculate the annual increases on the salaries in question 5
following a salary rise of 8%.

7   Find the overtime rates of pay at 'time and a half' corresponding to flat rates of 36p,   40p,   50p,   56p,   70p   and
85p per hour.

8 Repeat question 7 for 'time and a quarter'.

9 Calculate the gross week's pay for the following shop assistants who work a 42 hour week and are paid time and a quarter for overtime.

Assistant	Hourly rate	Hours worked
Annie Andrews	28p	42
Betty Bouncer	36p	48
Carole Carr	32p	45
Dorothy Dreadnought	30p	46

10 Find the amounts paid to a Health Club by each of the four girls in question 9, if the contribution was 2% of their gross wages. Answers to the nearest penny.

11 The Nimble Thimble Sewing Company pay their sales representatives a basic £780 a year with a 2% commission on the first £1 000 of monthly sales and $2\frac{1}{2}$% on the remainder.

Calculate the monthly salaries for representatives selling goods of value £800, £960, £1 400, £1 600 and £2 050 in a month.

12 Under the conditions of question 11, what sales are required to give a monthly salary of £80, £110 and £120?

13 Representatives of the 'Krazee Kard Ko' receive a basic salary of £1 020 a year with a $1\frac{1}{2}$% commission on sales. Calculate the gross monthly pay of representatives with sales of £840, £920, £990 and £1 240.

14 The car allowance for the representatives in question 13 is 3·2p per kilometre. Calculate the allowances for distances of 1 250, 2 680, 3 340 and 3 660 km.

15 An agent receives 2% commission on all goods sold. Find the sales to give a commission of £54.

Find the corresponding sales at $2\frac{1}{4}$% commission.

16 The Gruesome Mask Manufacturing Company pay their moulders a basic 24p per hour plus a bonus of $1\frac{1}{2}$p for Red Demons (R), 2p for Draculas (D) and $2\frac{1}{2}$p for Toothless Hags (T). Calculate the day's pay for a moulder working $7\frac{1}{2}$ hours, who produced 12R, 15D and 6T.

17 Repeat question 18 for (a) 14R, 9D and 16T, (b) 8R, 15D and 24T.

18 A Laboratory Assistant receives a starting salary of £720 with annual increments of £84. Calculate his monthly salary in the first and second years.

Calculate his monthly salary in the third year if there is a 10% pay rise.

19 'Toppers' – the hat manufacturers – arrange new piece work rates for their machinists. There is a basic 22p per hour plus a 2·2p per hat. Calculate the earnings, on a 7½ hour day, of girls who machined 30, 34, 37 and 44 hats.

20 'Toppers' pay time and a half on the basic rate for overtime. How much does a machinist who works 9½ hours producing 55 hats receive?

21 Salary scheme A offers a starting salary of £900 with 5 annual increments of £100. Salary scheme B also starts at £900 but has 5 annual increments of 10%.
   (a) Which scheme gives the higher salary after 2 increments? (b) Which is higher after 5 increments? (c) Which gives the greater total income over the 5 years? (d) Which scheme would be more profitable in the next 5 years?

HIRE PURCHASE

22 The cash price of a second hand motor cycle is £85. Hire Purchase calls for a deposit of 10% and 24 monthly payments of £3·75 each. Find (a) the deposit, (b) the total payment on HP, (c) the saving by cash purchase.

23 A bicycle sells for £38 cash. HP terms are a deposit of £3·80 followed by 12 monthly payments of £3·10. Find the difference between the two costs.
   The bicycle saves its purchaser 28p per week in bus fares for 50 working weeks. However, it depreciates by 20% in a year. Find the overall gain or loss if the cycle was bought on HP.

24 A cash discount of 5% is allowed on a £35 refrigerator. If the same refrigerator is bought on HP then there is a deposit of 20% followed by 18 monthly payments of £1·65.
   Find (a) the cash price, (b) the deposit for HP, (c) the total payment by HP, (d) the difference in the two costs.

25 Calculate the total HP payment on the goods listed below.

Goods	Cash price	Deposit	Payments
Skating boots	£12·40	20%	12 of £0·95
Cooker	£68	10%	24 of £2·75
Camera	£48	20%	18 of £2·25
Skin diving gear	£120	15%	24 of £4·60.

HOUSING

26 A Building Society offers Mr Hopeful an 80% mortgage on a £4 600 house. Surveyors fees are ½% of the value of the house, legal expenses are £96 and removal costs £28. Find (a) the amount of the Society's loan, (b) the surveyor's fee, (c) the

total amount of cash Mr Hopeful must find, (d) the cost of house insurance at $7\frac{1}{2}$p per £100.

27 The Backwater Building Society are prepared to make an 85% loan on a house of value £6 400. They specify that the loan must not exceed three times the client's gross income. Find (a) the client's minimum income if he takes the loan, (b) his cash contribution towards the house, (c) his further expenses if these are $6\frac{1}{2}$% of the purchase price.

28 A house has a rateable value of £94. Calculate (a) the contribution the owner makes to street lighting if this represents a rate of 0·8p in the £, (b) the extra payment he has to make if the rate is increased by 6p, (c) his half yearly payment of rates at 76p in the £.

29 A country town has a total rateable values of £224 000. What is (a) the product of 1p rate, (b) the total income from a rate of 84p, (c) the rate increase needed to pay for a £5 600 Health Centre, (d) the cost to an owner of a house of rateable value £108 of the new Centre?

30 The water rate for Knutsford is 12p in the £ and is paid in two half yearly instalments. Calculate the half yearly payment on houses of rateable value £70, £76, £85, £102 and £116.

INCOME TAX

Refer to the simplified tax table below for the remaining questions in this exercise.

TAX TABLE

Single person allowance £595, *Married man's allowance* £775
Children's allowances: not over 11 £200
                             12 to not over 16 £235
                             over 16, in full time education £265.
Income Tax Rate 30%.

31 Calculate the tax paid on £400 of taxable income.

32 Calculate the tax paid on £640, £880, £1 060 and £1 750 of taxable income.

33 A married man, with 2 children under 11, has a gross income of £1 960. Calculate (a) his total allowances, (b) his taxable income, (c) the amount of tax he pays.

34 Calculate the tax payable by (a) a single man with income of £920, (b) a single man with income of £1 890, (c) a married

man with one child of 13 and an income of £2 350, (d) a married man with two children under 12 and an income of £1 980, (e) a married man with four children aged 7, 8, 13 and a sixth former of 17, and an income of £2 430.

Calculating

Exercise 7

1 Give the value of 777, 835, 9 121, 484, 1 040, 676, 1 881 and 35·5 (a) to the nearest 100, (b) to the nearest 50.

2 (a) Write the following correct to 2 decimal places. 13·636, 5·749, 0·08646, 359·953, 67·276, 183·381 and 555·555. (b) Write the above numbers to 3 significant figures.

3 Express 506·8363 (a) correct to 3, 2 and 1 decimal places, (b) to 3, 2 and 1 significant figures.

4 (a) Write correct to 2 decimal places: 1·727, 0·1727, 0·01727, 0·001727. (b) Write the same numbers to 2 significant figures.

5 $\pi \simeq 3\cdot1416$. Express π correct to 2 then 3 decimal places and to 3, 2 and 1 significant figure.

6 $n = 12 \pm 0\cdot5$. Find the range of values of $3n$, $n + 5$, $n - 7$, $2n - 6$, n^2, $(n + 2)^2$.

7 If $x = 24 \pm 0\cdot5$ and $y = 20 \pm 0\cdot5$, find the range of values of $x + y$, $x - y$, $3x + 2y$, $3x - 2y$, xy.

8 Make a rough estimate of the value of $3\,030 \div 151$, $1\,080 \div 56$, $183\cdot7 \div 0\cdot51$, $5\,242 \div 0\cdot66$, $843\,000 \div 1\,208$.

9 Estimate the value of 100π, 300π, 36π, 70π, $\pi(8^2)$, $\pi(15\cdot6^2)$, $22\cdot2\pi$.

10 Give a quick estimate of $49 \times 32\cdot3$, $503 \times 68\cdot7$, $\dfrac{382 \times 15\cdot1}{29\cdot4}$, $\dfrac{1\,402}{39 \times 50\cdot07}$, $\dfrac{19\cdot3\pi}{12\cdot8}$, $\dfrac{7\cdot92 \times 101\cdot3}{(6\cdot95)^2}$

11 Use tables or slide rules to obtain more precise answers to questions 8, 9 and 10.

12 Estimate and then use square tables or slide rules to find the values of: (a) $15\cdot3^2$, $1\cdot53^2$, 153^2, $0\cdot153^2$, (b) $2\cdot84^2$, $38\cdot4^2$, 484^2, $0\cdot584^2$, (c) $11\cdot1^2$, 222^2, $3\cdot33^2$, $0\cdot444^2$, $0\cdot0555^2$.

13 Estimate and calculate using slide rule or square root tables: (a) $\sqrt{57\cdot7}$, $\sqrt{5\cdot77}$, $\sqrt{577}$, $\sqrt{0\cdot577}$. (b) $\sqrt{16\cdot1}$, $\sqrt{26\cdot2}$, $\sqrt{363}$, $\sqrt{0\cdot464}$, $\sqrt{0\cdot0565}$.

14 Evaluate: $\sqrt{16\cdot6^2 + 14\cdot4^2}$, $\sqrt{93^2 + 42^2}$, $\sqrt{6\cdot5^2 + \pi^2}$, $\sqrt{0\cdot18^2 + 0\cdot38^2}$.

15 Estimate and calculate using tables or slide rules,
(a) $\dfrac{1}{2\cdot3}$ $\dfrac{1}{23}$ $\dfrac{1}{0\cdot23}$ $\dfrac{1}{0\cdot0023}$ (b) $\dfrac{1}{15\cdot6}$ $\dfrac{1}{1\cdot56}$ $\dfrac{1}{0\cdot156}$ $\dfrac{1}{156}$
(c) $\dfrac{1}{4\cdot4} + \dfrac{1}{0\cdot44}$ $\dfrac{1}{6\cdot6} - \dfrac{1}{66}$ $\dfrac{1}{8\cdot35} - \dfrac{1}{83\cdot5}$ (d) $\dfrac{1}{2} + \dfrac{1}{2\cdot2} + \dfrac{1}{0\cdot22}$.

16 Arrange in order of increasing size: $\dfrac{1}{77}$ $\dfrac{1}{7}$ $\dfrac{1}{0\cdot77}$ $\dfrac{1}{7\cdot7}$.

17 Write down the logarithms to the base ten of 18·6, 279, 5·68, 0·164, 2 870, 0·0729, 0·00847.

18 Find the values of $10^{0\cdot432}$, $10^{1\cdot313}$, $10^{2\cdot505}$, $10^{\bar{2}\cdot505}$, $10^{3\cdot262}$, $10^{\bar{3}\cdot262}$, $10^{0\cdot262}$.

19 Find the numbers whose logarithms to the base ten are:
(a) 0·315, 2·315, $\bar{2}$·315, 4·315, (b) 2·666, $\bar{1}$·666, 3·666, $\bar{3}$·666, (c) 1·525, $\bar{2}$·525, 2·525, $\bar{6}$·525.

20 Simplify: (a) $0\cdot8 + \bar{1}\cdot6$, $0\cdot8 - \bar{1}\cdot6$, $\bar{1}\cdot6 \times 2$, $\bar{1}\cdot6 \div 2$,
(b) $0\cdot4 + \bar{1}\cdot7$, $0\cdot4 - \bar{1}\cdot7$, $\bar{1}\cdot7 \times 3$, $\bar{1}\cdot7 \div 3$,
(c) $\bar{1}\cdot5 + \bar{2}\cdot3$, $\bar{1}\cdot5 - \bar{2}\cdot3$, $\bar{1}\cdot5 \times 3$, $\bar{1}\cdot5 \div 2$,
(d) $\bar{1}\cdot5 + \bar{3}\cdot8$, $\bar{1}\cdot5 - \bar{3}\cdot8$, $\bar{1}\cdot5 \times 4$, $\bar{3}\cdot8 \div 2$.

21 Estimate, and use tables or slide rules to calculate:
(a) 476×231, (b) $476 \div 231$, (c) $231 \div 476$,
(d) $85\cdot3 \times 14\cdot7$, (e) $125 \div 34\cdot3$, (f) $1\cdot71 \times 15\cdot9$,
(g) $0\cdot832 \times 0\cdot571$, (h) $0\cdot832 \div 0\cdot571$, (i) $8\cdot32 \div 0\cdot571$,
(j) $\sqrt[3]{39\cdot7}$, (k) $\sqrt[3]{397}$, (l) $\sqrt[3]{0\cdot397}$, (m) $\sqrt[3]{1560}$,
(n) $\sqrt[3]{18\cdot2}$, (o) $\sqrt[4]{0\cdot182}$, (p) $15\cdot6^3$, (q) 156^3,
(r) $0\cdot156^3$, (s) $18\cdot3 \times 16\cdot4 \div 14\cdot6$, (t) $18\cdot3 \div (16\cdot4 \times 14\cdot6)$,
(u) $6/15\cdot8$, (v) $6/49$, (w) $6/0\cdot49$, (x) $\pi(15\cdot6)^2$,
(y) $\pi \times 15\cdot2 \div 18\cdot4$, (z) $\pi \times (15\cdot2)^2 \times 16$.

22 If $p = 48\cdot7$ and $q = 23\cdot6$, evaluate: pq, p/q, p^2, $\sqrt{q}$ $1/p + 1/q$.

23 When $r = 25\cdot6$ and $s = 5\cdot38$ find the value of rs, r/s, s^2 $\sqrt{r}$, $1/r + 2/s$.

24 $l = 19\cdot7$ and $m = 0\cdot216$. Arrange in order of increasing size: lm, m/l, l/m, l^2, $1/m$.

Use slide rules for the following questions.

25 1 inch ≃ 2·54 centimetres. Convert the following high jump records into centimetres. 3 ft 7 in, 4 ft 2 in, 4 ft 11 in, 5 ft 4 in, 5 ft 7 in and 5 ft 10 in.

26 (a) A fitter is paid 43p per hour. Calculate his pay for 4½, 5¼, 6¾ and 8½ hours. (b) Calculate the payment for overtime, at 'time and a half', for the same hours.

27 P.I.P. Petrol is priced at 37p per gallon. Find the cost of 3, 4½, 5, 5½ and 6 gallons.

28 Use 1 litre = 0·220 gallons and make a table of costs of 1, 2, 3 . . . . . . 10 litres of P.I.P. Petrol.

29 Use the formulae $C = 2\pi r$ and $A = \pi r^2$, to find the circumference and area of the circle with radius: 8·5, 11·2, 28·4, 6·06, 0·606 and 48 cm.

30 Compost Farm has fields of area 5·7, 6·3, 7·8, 8·5, 13·2 and 15·5 acres. Convert these areas to hectares, if 1 hectare = 0·405 acres.

Ratio and Proportion

Exercise 8

1 Find, in simplest form, the following ratios:
(a) 15 cm:2·5 m, (b) 18 h:3 days, (c) 35 degrees:1 right angle,
(d) 10° 18′:46° 21′, (e) £1·25:£2·75, (f) 17½p:£1·40,
(g) Days in September:Days in a leap year,
(h) Days in June: Days in a non-leap year.

2 Divide the 60 minutes of an hour in the ratio: (a) 3:2:1,
(b) 5:3:4, (c) 6:2:7, (d) 9:7:4.

3 The sides of a triangle are in the ratio 3:5:7. Find the lengths of the sides if the perimeter is 90 cm.
    Are the angles in the ratio 3:5:7?

4 The vital statistics of 'Metric Mary', in millimetres, are 900, 630, 990. Reduce her proportions to their simplest form.
    Ten years from now it is estimated that these will increase in the ratio 4:3, 10:7 and 15:11 respectively. What will then be her proportions?

5 Share £150 in the ratio (a) 3:7, (b) 8:7, (c) 8:17,
(d) 11:9, (e) 6:5:4, (f) 8:5:7, (g) 13:8:4.

6 A young athlete runs 400 m in 72 seconds and 5 000 m in 24 minutes. Find (a) the ratio of the distances, (b) the ratio of the times.

7 Jane weighed 45 kg. June weighed 70 kg. After two weeks of dieting Jane and June weighed 42 kg and 63 kg respectively. Find the ratio of (a) their original weights, (b) their new weights, (c) the reductions in their weights, (d) the proportionate reductions in their weights.

8 Concrete is made by mixing sand, cement and chippings in the ratio $2:1:5$. Find how much of each is needed for $\frac{1}{2}$ tonne of concrete. (1 tonne $= 1\,000$ kg).

9 Repeat question 8 for (a) $3:2:7$ and $2\cdot4$ tonne, (b) $3:1:4$ and $1\cdot6$ tonne, (c) $4:3:10$ and $3\cdot4$ tonne.

10 Calculate the angles of a triangle if these are in the ratio: (a) $1:3:5$, (b) $3:4:5$, (c) $9:10:11$, (d) $10:5:3$, (e) $8:13:15$.

11 A competition organiser decides to share the prize money of £2 400 in the ratio $20:5:3:2$. What is (a) the top prize, (b) the third prize, (c) the difference between the first and fourth prizes?

12 Three sons share an inheritance in the ratio $7:6:5$. The brother with the largest share divides this in the ratio $3:2:1$ between himself, his wife and his daughter. The daughter received £350. What was the original total inheritance?

13 Three pirates shared 8 500 pieces of eight in the ratio $9:5:3$. What was (a) the smallest share, (b) the difference between the two larger shares?

14 At the assizes, the pirates received sentences in the same ratio, $9:5:3$. The shortest sentence was $4\frac{1}{2}$ years. What were the other two sentences?

15 A machine prints 24 000 newspapers in 72 minutes. How long will it take to print 33 000 at this rate?

16 At 50 km/h a journey takes 54 minutes. How long will it take at 60 km/h? What speed is needed to complete the journey in 40 minutes?

17 At a barbecue 150 guests ate 360 'hot dogs'. At the next barbecue 500 'hot dogs' were provided for 200 guests. Assuming the same rate of consumption, how many were left over?

18 A canning machine fills 1 200 tins of strawberries per hour and takes 75 minutes to complete an order. How long would it

take at 1 500 tins per hour? What rate is needed to complete such an order in 48 minutes?

19   A clock takes 5 seconds to strike 3 o'clock. How long will it take to strike 6 o'clock?

20   The least time a train takes to pass four kilometre posts is 5 minutes. At this speed, what is its least time to pass seven kilometre posts?

21   In an office a machine opens 120 letters per minute and clears the morning mail in 18 minutes. How long will a new model, working at 150 per minute take?
If the mail increases by 40%, what speed is needed to clear the mail in the original 18 minutes?

22   The measurements of the sides of five rectangles are listed below. Sort these rectangles into sets of similarly shaped figures.
(a) 95 cm × 38 cm,   (b) 1 m × 40 cm,   (c) 550 m × 200 m,
(d) 0·8 m × 0·32 m,   (e) 5·5 m × 2 m.

23   The edges of two cubes are 6 cm and 10 cm respectively. State the ratio of   (a) the areas of their faces,   (b) their volumes, (c) the lengths of their diagonals.

24   The sides of two squares are 18 cm and 24 cm respectively. What is the ratio of   (a) their perimeters,   (b) their areas, (c) the volumes of the cylinders formed by rotating the squares through 360 degrees about one edge.

25   Calculate the missing scales in the table below.

Length scale	1:10	1:6	3:20		
Area scale				16:2500	4:49
Volume scale			64:125		

26   Find   (a) the length on the land in kilometres corresponding to 1 cm on a map,   (b) the area on the map in square centimetres corresponding to 1 km² on the land for maps with scales of   1:10 000;   1:25 000;   1:250 000;   1:1 000 000.

27   A colour transparency 36 mm by 24 mm is projected to give a picture 2·4 m in length. Calculate   (a) the width of the picture,   (b) the ratio of the area of the picture to that of the transparency.

28   A model aircraft is made to 1/50th scale – for lengths. What is the scale for   (a) the wing area,   (b) the height of the tail fin,   (c) the capacity of the fuel tank,   (d) the diameter of the wheels,   (e) the space in the cockpit?

29   A giant detergent bottle is made for advertising. It is 25 times as tall as the normal bottle. If the normal bottle holds

42 cm³ of detergent, how much does the giant bottle hold? What is the ratio of the areas of the labels on the two bottles?

30 A model is made of a new playing field to a scale of 1 : 100. What actual length corresponds to 4·8 cm on the model? What actual area corresponds to 9 cm² on the model? The total area of the field is 1·2 hectares. What is the total area of the model?

One of the soccer pitches on on the model has a slope of 1 in 30. What will be the slope of the actual pitch?

The miniature long jump contains 4 cm³ of sand. What is the corresponding volume for the full sized pit?

Trigonometrical Ratios (0-90°)

Exercise 9

1 Copy and complete the table opposite, giving values to two decimal places.

x	0°	30°	45°	60°	90°
sin x			·71		
cos x					0
tan x		·58			

2 Use your table from question 1 to sketch graphs, from $x = 0°$ to $90°$, of $y = \sin x$, $y = \cos x$ and $y = \tan x$. Why is it not possible to complete the graph for tan x?

3 Answer the following questions for acute angles. (a) Which ratios increase as the angle increases? (b) Why are mean differences for cosine subtracted? (c) Which ratios lie between 0 and 1? (d) Is it possible to have tan $x = 1\,000$? (e) What is the range of tan x?

4 Plot, on the same axes, graphs of $y = \sin x$ and $y = \cos x$, for x from 0° to 90°. Use your graphs to answer: (a) For what value of x does $\sin x = \cos x$? (b) Which ratio increases as x increases? What does the other ratio do? (c) Does $\sin x = \cos (90° - x)$? (d) Explain how your graph shows that $\sin 60°$ is not equal to $2 \sin 30°$. (e) Explain graphically why $\cos 45°$ is not equal to $\frac{1}{2} \cos 90°$.

5 Use tables to find sin x and tan x for x equal to $28° 28'$; $35° 53'$; $44° 44'$; $64° 46'$; $75° 57'$; $82° 28'$. Is tan $x >$ sin x for each of these angles?

6 Find cos x and log cos x for the angles given in question 5. Does log (cos x) – i.e. looking up cos x and then looking up the log of this number – give the same results?

7 Which of the following statements are impossible?
(a) sin $x = 1 \cdot 600$, (b) log cos $x = 0 \cdot 5131$, (c) tan $x = 18 \cdot 6$,
(d) cos $x = 4/3$, (e) sin $x >$ tan x, (f) log sin $x = \frac{1}{2}$.

8 Use tables to find angle x if (a) sin x, (b) cos x and
(c) tan x equals: $0 \cdot 1815$; $0 \cdot 3443$; $0 \cdot 6248$; $0 \cdot 6789$; $0 \cdot 7171$.

9 In Fig. 9.1, calculate:
(a) sides LN and NM given that LM $= 6$ cm and angle M $= 41°$
(b) sides LN and NM given that LM $= 8$ cm and angle L $= 52°$
(c) sides LN and LM given that MN $= 10$ cm and angle M $= 32°$
(d) sides LN and LM given that MN $= 9$ cm and angle L $= 63°$
(e) angle M given that LN $= 4$ cm and LM $= 10$ cm
(f) angle M given that NM $= 8$ cm and LM $= 12$ cm
(g) angle L given that LN $= 6$ cm and MN $= 9$ cm.

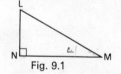

Fig. 9.1

Fig. 9.2

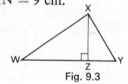

Fig. 9.3

10 In Fig. 9.2 calculate
(a) side RQ if PR $=$ PQ $= 8$ cm and angle P $= 54°$
(b) side PR if RQ $= 10$ cm and angle P $= 48°$
(c) angle P if PR $=$ PQ $= 9$ cm and RQ $= 4$ cm
(d) angle R if PR $=$ PQ $= 6$ cm and RQ $= 5$ cm
(e) the area of the triangle if RQ $= 6$ cm and angle P $= 36°$
(f) the area of the triangle if RQ $= 8$ cm and angle R $= 66°$.

11 In Fig. 9.3 calculate
(a) sides XZ and YZ if WX $= 6$ cm, angle W $= 34°$ and angle X $= 100°$
(b) sides XZ and XY if WZ $= 8$ cm, angle W $= 38°$, and angle Y $= 68°$
(c) sides WZ, XZ and angle Y if WX $= 9$ cm, WY $= 15$ cm and angle W $= 41°$
(d) the area of triangle WXY if WX $= 8$ cm, angle W $= 27°$ and angle Y $= 58°$.

12   P, Q, R and S (Fig. 9.4) are four corners of a regular nonagon (nine-sided figure). The nonagon is drawn in a circle,
centre O and radius 5 cm.   (a) How many triangles congruent

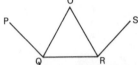

Fig. 9.4

to triangle ROQ are there in the nonagon?   (b) How many
angles are there equal to angle ROQ?   (c) What is the value of
angles ROQ, POR, POS?   (d) What can be said about lengths
OP, OQ, OR and OS?   (e) What type of triangle are triangles
OPQ, OQR and ORS?   (f) Calculate the length RQ.   (g) Calculate the lengths PR and PS.

13   Copy and complete the following table of ratios for acute
angles.

$\sin x$	$\frac{3}{5}$					$\frac{9}{41}$
$\cos x$		$\frac{5}{13}$		$\frac{21}{29}$		
$\tan x$			$\frac{8}{15}$		$\frac{7}{24}$	

14   Compile a complete version of the table below.

$x =$	0°	30°	45°	60°	90°
$\sin x =$			$\dfrac{1}{\sqrt{2}}$		
$\cos x =$	1				
$\tan x =$				$\sqrt{3}$	

15   Use the table from question 14 to calculate the missing
lengths in Fig. 9.5. Leave any square roots in your answers.

	WX	WY	YZ	YZ	WZ
(a)	20 m				
(b)		6 m			
(c)				5 m	
(d)			6 m		
(e)					$7\sqrt{3}$ m

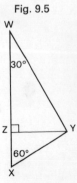

Fig. 9.5

16   A ship sails on a course of 058° for 5 km. How far has it travelled North and how far East from its starting point?

17   Repeat question 16 for   (a) 046° and 8 km,   (b) 076° and 9 km,   (c) 085° and 10 km.

18   In what direction did a yacht sail if its course took it 5 km East and 6 km South of its starting point?

19   Repeat question 18 for   (a) 3 km East and 5 km South   (b) 4 km East and 6 km North   (c) 6 km West and 5 km North   (d) 5 km West and 10 km South.

20   A light aircraft flies 80 km due East then a further 80 km in a direction 050°. How far is it then from its starting point?

21   A straight road climbs a hillside at a steady 15° to the horizontal. A car travels 840 m up the road. How far does the car (a) rise vertically,   (b) travel horizontally? Answers to nearest metre.

22   An aerial photograph was taken when the angle of elevation of the sun was 54°. Calculate the heights above their surroundings of   (a) a dam with a shadow of length 80 m,   (b) a tower with a shadow of length 46 m,   (c) a hill with a shadow of length 780 m.

23   Sprinters S_1 and S_2 leave position P at the same time. S_1 runs due South at 8 m/s and S_2 runs due East at 9 m/s. Calculate the direction $S_1 S_2$ after 4 seconds. If they keep up the same speeds what will the direction $S_1 S_2$ be after 6 seconds?

24   The water-chute at a fun fair is 30 m high and is inclined at 40° to the horizontal. Calculate   (a) the horizontal and   (b) the inclined lengths of the chute.

25   An exuberant youth slides down some banisters.   These are inclined at 35° to the horizontal and are 5 m long. Calculate how far he moves horizontally and how far vertically.

26   A rectangle has sides 15 cm and 6 cm. Calculate the acute angle between its diagonals.

27   The diagonals of a rectangle are of length 16 cm. The angle between the diagonals is 68°. Calculate the sides of the rectangle.

28   A flagpole 30 m high stands at one corner of a rectangular parade ground with sides 120 m by 90 m. Calculate the angles of elevation of the top of the pole from the other three corners of the parade ground.

29   The jib of a crane can be elevated between 25° and 75°. The jib is 40 m long and its hook hangs vertically. Calculate the

distance from the base of the jib of the furthest and closest points that can be reached by the hook.

30   A rope 12 m long hangs from the branch of a tree. A monkey swings on the rope 15° either side of the vertical. How far does the monkey travel horizontally in one complete swing? (i.e. out and back to his starting point).

Real Numbers

Exercise 10

1   Simplify: $(+15) \div (+3)$, $(+15) \div (-3)$, $(-15) \div (+3)$, $(-15) \div (-3)$; $(+8) \times (+4)$, $(+8) \times (-4)$, $(-8) \times (+4)$, $(-8) \times (-4)$; $(+18) \div (+6)$, $(+18) \div (-6)$, $(-18) \div (+6)$, $(-18) \div (-6)$; $(+7) \times (+5)$, $(-7) \times (+5)$, $(+7) \times (-5)$, $(-7) \times (-5)$; $(+6) + (+3)$, $(+6) + (-3)$, $(+6) - (+3)$, $(+6) - (-3)$; $(+12) + (+8)$, $(+12) + (-8)$, $(+12) - (+8)$, $(+12) - (-8)$.

2   If $x = -3$, $y = -2$ and $z = -5$, find the value of: x^2, x^3, xy, yz, xyz, $x - y + z$, $x - y - z$, xy/z, xy^2/z.

3   If $p = -4$, $q = -3$ and $r = 6$, find the value of: $p + q + r$, pqr, p^2, q^3, $(p + q)/r$, $q/(p + r)$, $pq + r$, p^2r, q^2r.

4   (a) Find the value of $2x^2 - 3x + 4$ when $x = -2$, -5 and -7. (b) Evaluate $3x^2 + 6x - 5$ when $x = -3$, -7, -9 and -11.

5   Evaluate   (a) $\dfrac{r + s}{rs}$,   (b) $\dfrac{r - s}{rs}$,   (c) $\dfrac{s - r}{sr}$,

(d) $\dfrac{r^2 + s^2}{r - s}$ when $r = -2$ and $s = -3$.

6   Simplify the following: $3 - (x + 2)$, $3 - (x - 2)$, $2(x - 5) - 2(x - 3)$, $4(x - 2) - 3(x - 5)$, $4x - 2(3x - 6)$, $5x - 3(4 - x)$, $2x^2 - 3x(x - 5)$, $4x^2 - 7x(5 - x)$, $2\{x - 3(x - 5)\}$, $6\{x - 4(3 - x)\}$.

7   Copy Fig. 10.1 and continue each pattern up to six triangles.

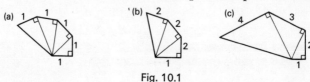

Fig. 10.1

Calculate the lengths of the sides opposite the right angles. Leave your answers as square roots, if they do not simplify further.

8   Calculate the lengths of the diagonals of the rectangles with sides   4 m × 5 m,   3 m × 8 m,   6 m × 2 m,   7 m × 9 m, 11 m × 2 m,   and   8 m × 4 m. Leave your answers in simplest root form.

9   Find the areas of the squares with sides: 3 cm, $\sqrt{3}$ cm, $3\sqrt{3}$ cm;   7 cm, $\sqrt{7}$ cm, $2\sqrt{7}$ cm;   13 cm, $\sqrt{13}$ cm, $5\sqrt{13}$ cm.

10   Find the sides of the squares with areas: 16 cm², 61 cm²,   25 cm²,   52 cm²,   36 cm²,   63 cm,   81 cm², 18 cm².

11   Calculate the radius of the circles with areas: 2π cm², 8π cm², 5π cm², 20π cm², 7π cm², 63π cm², 98π cm², 242π cm².

12   Find and correct any errors in the following statements. $\sqrt{900} = 30$, $\sqrt{9\,000} = 300$, $\sqrt{0\cdot09} = 0\cdot3$; $\sqrt{16\cdot9} = 1\cdot3$, $\sqrt{169} = 13$, $\sqrt{0\cdot169} = 0\cdot13$; $\sqrt{81} = 9$, $\sqrt{81\,000} = 900$, $\sqrt{0\cdot81} = 0\cdot9$.

13   In each of the following sets find the one number that is not equal to the other three.
(a) $\{2\sqrt{8},\quad \sqrt{32},\quad 4\sqrt{2},\quad \sqrt{48}\}$,
(b) $\{\sqrt{75},\quad 3\sqrt{5},\quad 5\sqrt{3},\quad \tfrac{1}{2}\sqrt{300}\}$,
(c) $\{\sqrt{27},\quad \tfrac{1}{2}\sqrt{108},\quad 3\sqrt{9},\quad 3\sqrt{3}\}$,
(d) $\{\sqrt{125},\quad 5\sqrt{5},\quad 5\sqrt{25},\quad \tfrac{1}{2}\sqrt{500}\}$.

14   Solve the equations, giving the answers as roots in their simplest form: $x^2 = 5$, $x^2 = 72$, $x^2 + 6 = 58$, $x^2 - 4 = 24$, $x^2 - 9 = 41$, $x^2 + 9 = 41$.

15   Solve: $x^2 + 5x + 1 = 0$, $x^2 + 5x - 1 = 0$, $x^2 - 3x + 1 = 0$, $2x^2 - 6x - 2 = 0$, leaving square roots in your answers.

16   Express in root form: sin 45°,   cos 30°,   tan 60°, cos 45°,   tan 30°.

17   In triangle WXY, angle X = 90°, angle Y = 60°, length WX = 16 cm. Calculate, leaving roots in your answers, lengths XY and WY and the area of the triangle.

18 Figure 10.2 illustrates a rectangle with sides $8\sqrt{2}$ and 8 cm. Calculate the sides of the half, quarter and one eighth rectangles. Show that the whole family of rectangles has sides in the ratio $\sqrt{2}:1$.

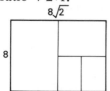

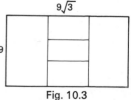

Fig. 10.2   Fig. 10.3

19 Figure 10.3 illustrates a rectangle with sides $9\sqrt{3}$ and 9 cm. Calculate the sides of the third and one ninth rectangles. Show that this family of rectangles has sides in the ratio $\sqrt{3}:1$.

20 Rule a number line using a scale of 1 cm to 1 unit. Plot the approximate positions on the line of the numbers, 5, $\sqrt{5}$, 7, 11, 2·3, −3, −4·6, $5\frac{1}{2}$, $\sqrt{3}$, $\sqrt{7}$ and π.

21 Repeat question 20 for 2, $\sqrt{2}$, 4·8, −5, $\frac{1}{2}\pi$, $\sqrt{5}$, $-\sqrt{5}$, 4·3̇3̇, $-4\frac{3}{4}$ and $\sqrt{12}$.

22 Say whether the following numbers are whole numbers, rational numbers, integers etc.
(a) 56, −56, 5/6, 5·6, 5·66̇6̇, 5·606006 . . .
(b) 38, $\sqrt{38}$, $(-38)^2$, 3/8, 3·8, 10^{38}
(c) 22/7, 3·14, π, $\sqrt{22}$, $22\sqrt{7}$, 22·7.

23 What set of numbers is obtained by (a) dividing pairs of natural numbers, (b) multiplying pairs of whole numbers, (c) multiplying pairs of negative integers, (d) subtracting pairs of whole numbers, (e) squaring pairs of whole numbers, (f) multiplying natural numbers by two, (g) multiplying the natural numbers by −1, (h) squaring the integers?

24 Say which set(s) of numbers you would use to (a) count sheep, (b) record decimal currency, (c) write percentages, (d) count the rungs on a ladder, (e) measure the length of the ladder, (f) record steps up or down the ladder from a rung half way up, (g) give the score in a football match, (h) record goal average.

25 Which set(s) of numbers are needed to solve: $2x = 12$, $2x = 13$, $2x + 1 = 7$, $2x + 7 = 1$, $x^2 = 9$, $x^2 = 10$, $x^2 - 16 = 0$, $x^2 - 15 = 0$?
What can be done about $x^2 + 15 = 0$?

26 What types of numbers are used in the following? (a) Five minutes to three. (b) A quarter past six. (c) The odds are 6 to

4 on. (d) 54%. (e) The length of the diagonal is $\sqrt{17}$ cm.
(f) A gradient of 1 in 5. (g) $\pi \simeq 3\cdot14$.

Modular Arithmetics

Exercise 11

1   The days of the week have an arithmetic modulo 7. Find the day that is   (a) 22 days after a Wednesday,   (b) 34 days after a Friday,   (c) 45 days after a Monday,   (d) 31 days before a Tuesday,   (e) 16 days before a Saturday.

2   The twelve hour clock has an arithmetic modulo 12. Use this to find what time it is   (a) 5h after 8 00 pm,   (b) 11 h after 4 00 pm,   (c) 7 h before 3 00 am,   (d) 18 h after 11 00 pm, (e) 25 h before 6 00 am.

3   Rotations, measured in degrees, form a system modulo 360. Find the angle equivalent to a rotation of $390°$, $470°$, $750°$, $1\,000°$, $1\,234°$, $2\,222°$, $35\,350°$.

4   Complete addition and multiplication tables for arithmetic modulo 4. Use your tables to find, where possible
(a) $3 + 2$, $1 + 3$, $1 + 2 + 3$, $3 + 2 + 1$,
$2 + 2 + 2$, 3×2, 2×2, $3 \times 2 \times 3$,
(b) $3 - 2$, $2 - 3$, $1 - 3$, $1 - 2$,
(c) $3 \div 2$, $2 \div 3, 1 \div 2$, $\frac{1}{3}$.

5   Use your tables from question 4 to solve, where possible, $2x \equiv 3$, $3x \equiv 1$, $x + 2 \equiv 3$, $x^2 \equiv 1$, $x^2 \equiv 0$.

6   Compile addition and multiplication tables for arithmetic modulo 5. Use your tables to find: (a) $2 + 4$, $3 + 4$, $1 + 2 + 4$, $2 + 3 + 4$,   (b) $4 - 2$, $2 - 4$, $1 - 3$, $(1 - 2) - 4$, $1 - (2 - 4)$,   (c) 3×2, 3×4, $2 \times 3 \times 4$, 4^2, 4^3,   (d) $2 \div 4$, $1 \div 3$, $4 \div 3$, $3 \div 4$, $\frac{3}{2}$, $\frac{2}{3}$, $\frac{1}{4} \div \frac{2}{3}$.

7   Use your tables from question 6 to solve, where possible: $2x \equiv 1$, $3x \equiv 2$, $x + 3 \equiv 2$, $x - 3 \equiv 4$,
$2x - 3 \equiv 4$, $x^2 \equiv 4$, $x^2 - 1 \equiv 0$, $x^2 + 3x + 2 \equiv 0$.

8   Make a table of values of x^2, x^3, x^4, for arithmetic modulo 5. Where possible, solve: $x^2 \equiv 1$, $x^2 \equiv 2$, $x^2 \equiv 4$, $x^3 \equiv 3$, $x^4 \equiv 2$, $x^4 \equiv 1$, $x^2 - 3 \equiv 2$, $x^3 + 2 \equiv 0$.

9   Write out addition and multiplication tables for arithmetic modulo 6.   Find: (a) $3 + 4$, $5 + 5$, $1 + 3 + 5$,

$2 + 3 + 4 + 5$, (b) 3×4, 5×5, $3 \times 3 \times 3$,
$2 \times 3 \times 4$, (c) 3^2, 4^2, 4^3, 3^4, (d) $3 - 5$, $2 - 4$,
$1 - 5$, $3 - (4 + 5)$, $(3 - 4) + 5$, (e) $2 \div 4$, $3 \div 3$,
$0 \div 3$, $5 \div 4$.

10   Using your tables from question 9, solve where possible:
$x + 4 \equiv 2$, $x - 5 \equiv 4$, $2x \equiv 4$, $3x \equiv 2$, $x^2 \equiv 4$,
$x^2 \equiv 1$, $x^2 - 5 \equiv 5$, $4x^2 \equiv 5$, $5x^2 \equiv 3$.

11   Spectators enter a football ground through a turnstile that
makes a quarter turn for each football fan who enters. The stile
turns clockwise and the arms point North, South, East and West.
Consider the arm that points North initially. In what direction
will it point when the total number of spectators passing
through is: (a) 1,   3,   4,   6,   9,   18,   22,   29,
(b) 305,   3 050,   3 600,   4 283,   6 151,   6 666?

12   A cistern is slowly filled with water. When the volume of
water reaches 4 litres it is automatically discharged leaving the
cistern empty. How much water remains in an originally empty
cistern if the following amounts are slowly poured in: (a) 31,
51,   71,   101,   121,   181,   4 001,   (b) 3 cans of 31 each, 5
cans of 21 each, 4 cans of 61 each?

13   A conjuror's table has a spring-loaded trap door. This door
will support five cubes but when a sixth cube is added all six fall
through. The trap then closes, leaving no cubes visible.
   Make an addition table for this situation. Use your table to
find how many cubes are visible if   (a) 8 cubes are added, one
by one,   (b) 11 cubes are added, one by one,   (c) 15 cubes are
added, one by one,   (d) 20 cubes are added in piles of 2,
(e) 20 cubes are added in piles of 4.

14   Ice gathers on a mountain peak. When the amount of ice
reaches 15 tonnes there is an avalanche and all the ice slides off.
At the end of one summer the peak was clear and during the
winter months the snowfall on the peak was equivalent to
53 tonnes of ice.   (a) How many avalanches were there?
(b) How much ice was left?   (c) How much more ice must form
to reduce this amount to 5 tonnes?   (d) What modular arithmetic is this?

Number Bases

Exercise 12

1   Write the denary numbers from 1 to 15 in binary form.

2 Express the following binary numbers in denary form: 1101, 1011, 1100, 10101, 111001, 101011, 110110110.

3 Convert from denary to binary form: 19, 23, 35, 41, 57, 69, 73, 192.

4 Write the following bicimals as denary fractions: 0·1, 0·011, 0·101, 0·111, 0·0101, 0·10101.

5 Change these denary fractions to bicimals: $\frac{1}{2}$, $\frac{3}{4}$, $\frac{1}{8}$, $\frac{1}{4}$, $\frac{5}{8}$, $\frac{9}{16}$, $\frac{7}{32}$.

6 Write in binary form the denary numbers: $15\frac{1}{2}$, $19\frac{1}{8}$, $23\frac{3}{4}$, $29\frac{3}{8}$, $37\frac{11}{16}$.

7 Express the following bicimals as mixed denary numbers: 101·11, 110·01, 1001·101, 1101·111, 1010·1001.

8 Carry out the following calculations, in binary form: (a) $1101 + 101$, $101 + 1101 + 111$, $1011 + 110$, $11011 + 1101$, (b) 1101×101, 1011×110, 11101×110, 10101×111, (c) $111 - 101$, $11001 - 101$, $100011 - 111$, $11110 - 101$, (d) 10^{10}, 110^{10}, 110^{11}, 111^{11}.

9 If $x = 11011$ and $y = 1001$ find, in binary form, $x + y$, $x - y$, xy, x/y, y^{10}.

10 Repeat question 9 for (a) $x = 111011$ and $y = 111$, (b) $x = 101000$ and $y = 1010$,   (c) $x = 10010$ and $y = 110$.

11 The numbers in the following equations are in binary form. Find the values of x.
(a) $101x + 11 = 10010$,   (b) $111x - 10011 = 11110$,
(c) $x^{10} + 101x + 110 = 0$,   (d) $x^{10} - 11x - 100 = 0$,
(e) $11^x = 1010001$.

12 Find the value of the number 123 if it is written to the base: four, five, six, eight. Why is it not in base two?

13 Arrange in descending order 222_8, 333_4, 444_5, 111_9.

14 Arrange in ascending order 201_3, 1011_2, 33_4, 43_8, 101_6.

15 Change to ascending order $0·33_4$, $0·22_5$, $0·46_8$.

16 In what bases, less than ten, are the following numbers prime? 13, 17, 21, 31, 42?

17 Consider the number 1011. (a) Write this as a denary number if it is in base 2, 5, 8. (b) Assume it is in denary form and write it in base 8, 5, 2. (c) Explain why 1011 is an odd number whatever its base. (d) Is 1111 an odd number for any base?

18   In the duo-decimal (base 12) system let 't' represent ten and 'e', eleven. Find the denary value of: 2te,   t3e,   et9,   5et, tte,   t42.

19   If $x = 321_4$ and $y = 23_4$, calculate $x + y$, $x - y$, xy, y^2, and the remainder when x is divided by y.

20   Repeat question 19 for x and y in base 5 and base 6.

21   (a) Say how to recognise if a binary number is divisible by: 2,   4,   8,   64.   (b) What special feature does a number to base three have if it is divisible by: 3,   9,   27,   243? (c) How can odd numbers be distinguished from even numbers in   (1) any even base,   (2) any odd base?

22   Find the bases of the numbers used in the following calculations.

$$
\begin{array}{ccccc}
121+ & 132+ & 132- & 114- & 34 \times 2 = 123_{7} \\
13 & 63 & 63 & 45 \\
\hline
200, & 225, & 36, & 25,
\end{array}
$$

$55 \times 21 = 1155,$ $222 \div 13 = 12,$ $3015 \div 15 = 201.$

Standard Form

Exercise 13

1   Express the populations of the following countries in standard form, to 2 significant figures.   (a) The United Kingdom   54 060 000,   (b) Gambia   333 000,   (c) Ceylon 10 600 000,   (d) USSR   226 000 000,   (e) Vatican City 940,   (f) Iceland   190 000,   (g) China   656 000 000.

2   Write the following areas in full.   (a) Sahara Desert 8.42×10^6 km², (b) Lake Ontario 1.95×10^4 km², (c) Bahrein 5.52×10^2 km²,   (d) Australia 7.7×10^6 km²,   (e) San Marino 5.95×10 km²,   (f) The Earth 5.6×10^8 km².
   Which of these areas are roughly comparable in size?

3 (a) Write in standard form the speed of light, 300 000 000 m/s. (b) Write in full the speed of sound, $3\cdot3 \times 10^4$ cm/s. What is this in m/s?

4 (a) Write in full, the average volume of water passing over Stanley Falls per minute is $1\cdot02 \times 10^6$ m^3. (b) Write, in standard form, 'The average volume of water passing over Niagara Falls per minute is 344 000 m^3.'

5 (a) The distance from the earth to the moon is $3\cdot85 \times 10^5$ km. Write this in full. (b) The distance from the earth to the sun is 149 000 000 km. Write this in standard form. (c) Write both distances both ways in metres.

6 The mass of the electron is approximately $9\cdot11 \times 10^{-31}$ kg. If this number is written out in full, how many zeros are there between the decimal point and the figure 9? Express the mass in grammes.

7 The volume of a particle is $1\cdot5 \times 10^{-23}$ cm^3. Write this in km^3 in standard form.

8 Calculate pq, q/p, p^2 and q^2, in standard form when
(a) $p = 5 \times 10^4$, $q = 6 \times 10^3$, (b) $p = 4 \times 10^5$,
$q = 3 \times 10^{-2}$, (c) $p = 6 \times 10^8$, $q = 9 \times 10^{-6}$,
(d) $p = 5 \times 10^{-4}$, $q = 7 \times 10^{-5}$.

9 Simplify, and give answers in standard form:

(a) $\dfrac{1\cdot8 \times 10^5 \times 4\cdot8 \times 10^7}{3\cdot6 \times 10^4}$ (b) $\dfrac{4\cdot5 \times 10^6 \times 1\cdot2 \times 10^4}{6 \times 10^{-5}}$

(c) $\dfrac{7\cdot2 \times 10^7}{1\cdot8 \times 10^5 \times 8 \times 10^{-4}}$ (d) $\dfrac{3\cdot5 \times 10^6 \times 2\cdot4 \times 10^8}{8\cdot4 \times 10^{-5}}$

10 Calculate, and express your answers in standard form:
(a) the number of matches in 1 500 000 boxes with an average of 44 matches per box, (b) the distance travelled by light in 15 minutes at 300 000 km/s, (c) the number of cubic centimetres in a cube of side 4 metres, (d) the thickness of a page if 960 pages are $3\cdot6$ cm thick – answer in cm, mm and m, (e) the volume of a speck of dust if 1 800 000 specks fill $2\cdot7$ cm^3 – answer in cm^3, mm^3, and m^3, (f) the number of hours in a lifetime of 'three score years and ten' – ignore leap years and answer to 2 significant figures.

The Algebra of Real Numbers

Exercise 14

1 Use figures 14.1a and 14.1b to illustrate
(a) $5(2x + 3) = 10x + 15$,
(b) $p(q + r - s) = pq + pr - ps$.

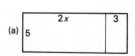 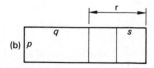

Fig. 14.1

2 Expand: $3(5a + 7)$; $4(a + 2b + 3c)$; $5(3r - 2s + t)$; $x(2x + 3)$; $3y(4y + 5)$; $2z(5 - 3z)$; $6s(s^2 - 3s + 4)$.

3 Use figures 14.2, a, b and c to expand
(a) $(2x + 3)(x + 1)$, (b) $(2x - 3)(x + 1)$,
(c) $(2x - 3)(x - 1)$.

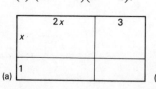 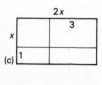

Fig. 14.2

4 Expand: $(x + 3)(x + 2)$; $(y + 6)(y - 3)$; $(z + 4)(z - 5)$; $(3p + 2)(p - 3)$; $(2q - 5)(q + 6)$; $(5r + 1)(r - 3)$; $(4s + 5)(2s - 7)$; $(3t + 8)(2t - 5)$.

5 Use figures 14.3a and 14.3b to obtain expansions for $(x + y)^2$ and $(x - y)^2$.

Fig. 14.3 (a)

6 Expand: $(x + 1)^2$; $(y - 2)^2$; $(z + 3)^2$; $(2p - 1)^2$;

$(4s + 3)^2$; $(5t - 7)^2$; $(a + b)^2$; $(a + 2b)^2$; $(2a - 3b)^2$;
$(5a - 4b)^2$; $(4a - 5b)^2$.

7 Draw illustrations and hence expand: $(a + b + c)^2$;
$(a + 2b + 3c)^2$; $(a + b - c)^2$; $(a - b + c)^2$.

8 Factorise and evaluate: (a) $pq + pr$,
$15 \times 8\cdot4 + 15 \times 1\cdot6$, (b) $pq + q^2$, $57 \times (23 + 23^2)$,
(c) $pq - pr$, $24\cdot2 \times 27\cdot6 - 24\cdot2 \times 2\cdot6$;
$14\cdot9 \times 27\cdot8 - 14\cdot9 \times 7\cdot8$.

9 Express in factor form: $x^2 + 2x$; $y^2 - 5y$;
$3z^2 + 4z$; $8p^2 + 4p$; $3q - 9q^2$; $4r^3 + 24r^2$;
$5s^2 - 20s^4$; $18t^3 + 12t^2$; $2r + \pi r$; $2\pi R + 2\pi r$;
$\pi r l + \pi r^2$; $2\pi r^2 + 2\pi r h$.

10 Factorise: $x^2 + 7x + 10$; $y^2 + 10y + 21$;
$z^2 + 13z + 36$; $p^2 - 4p - 12$; $q^2 - 5q - 24$;
$r^2 + 4r - 45$; $2s^2 + 7s + 5$; $3t^2 + 14t + 8$;
$5n^2 - 7n - 6$; $2p^2 + 7p + 3$; $6q^2 + 19q + 10$;
$8r^2 + 18r + 9$; $6a^2 - 5a - 6$; $8b^2 - 18b - 5$;
$6c^2 - c - 12$; $12d^2 + 17d - 5$; $x^2 + 6x + 9$;
$x^2 - 10x + 25$; $x^2 - 8x + 16$; $4x^2 + 4x + 1$;
$9x^2 - 12x + 4$; $n^2 - 9$; $q^2 - 121$; $4r^2 - 25$;
$9s^2 - 49$; $3x^2 - 12$; $2y^2 - 18$; $5z^2 - 80$.

11 Factorise $x^2 - y^2$ and evaluate $89^2 - 11^2$;
$33^2 - 17^2$; $(6\frac{1}{2})^2 - (3\frac{1}{2})^2$; $(7\cdot9)^2 - (2\cdot1)^2$;
$(17\cdot2)^2 - (2\cdot8)^2$; $(504)^2 - (204)^2$.

12 Factorise $\pi R^2 - \pi r^2$ and find the value of this expression
when $\pi = 22/7$, $R = 9$ and $r = 5$; $R = 17$ and $r = 11$;
$R = 23$ and $r = 12$.

13 Express as the product of two factors:
$ac + bc + ad + bd$; $x^3 + 2x^2 + 3x + 6$;
$pr + 2ps + qr + 2qs$; $rs + 3rt + 2s^2 + 6st$;
$pq - pr + q^2 - qr$; $4xy - y^2 + 8xz - 2yz$.

14 Complete the following to form sets of equivalent fractions
(a) $\dfrac{x}{y} = \dfrac{3x}{-} = \dfrac{}{8y} = \dfrac{-5x}{-} = \dfrac{x^2}{} = \dfrac{}{y^2}$.

(b) $\dfrac{p}{q} = \dfrac{6p}{-} = \dfrac{}{9q} = \dfrac{p^3}{-} = \dfrac{}{3q^2} = \dfrac{p(p + 1)}{} = \dfrac{}{q(q - 4)}$.

(c) $\dfrac{x - 3}{5} = \dfrac{}{10} = \dfrac{4x - 12}{} = \dfrac{}{5(x + 2)} = \dfrac{x^2 - 3x}{} = \dfrac{x^2 - 9}{}.$

(d) $\dfrac{r + s}{r - s} = \dfrac{3r + 3s}{} = \dfrac{}{s - r} = \dfrac{2(r + s)}{} = \dfrac{(r + s)(r - s)}{}.$

15 Express as single fractions: $\dfrac{1}{2x} + \dfrac{1}{3x}$; $\dfrac{1}{3x} - \dfrac{1}{4x}$;

$\dfrac{5}{6a} + \dfrac{1}{3a}$; $\dfrac{4}{9a} + \dfrac{1}{3a}$; $\dfrac{4}{9b} - \dfrac{1}{6b}$; $\dfrac{8}{15c} - \dfrac{1}{6c}$;

$\dfrac{4}{x} - \dfrac{3}{x}$; $\dfrac{7}{18x} + \dfrac{5}{8x^2}$.

16 Simplify: $\dfrac{1}{x} + \dfrac{1}{x+2}$; $\dfrac{1}{x} - \dfrac{1}{x+3}$; $\dfrac{1}{x-2} - \dfrac{1}{x}$;

$\dfrac{1}{x-3} - \dfrac{1}{2x}$.

17 Write as single fractions: (a) $\dfrac{1}{x+2} + \dfrac{1}{x+3}$,

(b) $\dfrac{1}{x+2} - \dfrac{1}{x+3}$, (c) $\dfrac{2}{x+4} - \dfrac{1}{x+3}$,

(d) $\dfrac{1}{x+1} - \dfrac{1}{2x+1}$, (e) $\dfrac{1}{x-2} - \dfrac{1}{2x-4}$,

(f) $\dfrac{1}{x+2} + \dfrac{2}{x+3}$, (g) $\dfrac{2}{5x+1} - \dfrac{1}{3x-1}$.

18 Simplify: (a) $\dfrac{x}{x^2-1} + \dfrac{1}{x-1}$, (b) $\dfrac{1}{x^2-9} - \dfrac{1}{x+3}$,

(c) $\dfrac{x}{x^2-1} - \dfrac{1}{x+1}$, (d) $\dfrac{1}{2x-1} + \dfrac{1}{(2x-1)^2}$,

(e) $\dfrac{1}{(x+3)^2} - \dfrac{1}{x^2-9}$.

19 Simplify: (a) $\dfrac{2xy}{4y^2}$; $\dfrac{4x^3}{10x^2}$; $\dfrac{15y^5}{10y^3}$; $\dfrac{10a^2b}{5ab}$

(b) $\dfrac{3r+6s}{r+2s}$; $\dfrac{ab-b^2}{2a-2b}$; $\dfrac{x-3}{3-x}$; $\dfrac{p-q}{2q-2p}$; $\dfrac{x+2}{x^2-4}$;

$\dfrac{s-t}{t^2-s^2}$

(c) $\dfrac{pq-q}{p^2-3p+2}$; $\dfrac{5x+5}{x^2-2x-3}$; $\dfrac{4s+8t}{s^2+4st+4t^2}$

(d) $\dfrac{8p^2q}{r} \div \dfrac{4pq}{r^2}$; $\dfrac{y-5}{y-4} \div \dfrac{10-2y}{3y-12}$; $\dfrac{x+2}{x^2-9} \div \dfrac{2x+4}{x-3}$.

The Solution of Equations

Exercise 15

Solve the following:

1. $x + 7 = 13.$ 2. $2x + 7 = 13.$ 3. $x - 7 = 13.$

4. $5x - 7 = 13.$ 5. $x + 3 = 15.$ 6. $4x + 3 = 15.$

7. $x - 3 = 15.$ 8. $6x - 3 = 15.$ 9. $4x + 8 = 2x + 15,$

10. $5x - 2 = 2x + 7.$ 11. $4x - 1 = 6x + 19.$

12. $5x + 3 = 24 - 2x.$ 13. $12 - 7x = 4x - 10.$

14. $6x - 5 = 27 - 2x.$ 15. $4(2x + 3) - 9 = 3(x + 6).$

16. $3(2x + 7) - 4(x - 5) = 50.$

17. $4(2x + 5) - 2(x - 6) = 8.$

18. $(1.2)(x + 4) = 8.4.$ 19. $\frac{1}{4}(x + 3) - \frac{1}{3}(x + 4) = 5.$

20. $\frac{1}{3}(5x + 2) - \frac{1}{2}(3x + 5) = 1.$

21. $\frac{1}{4}(2x + 1) = \frac{1}{3}(5x + 6).$

22 Find the three angles of a triangle if   (a) one angle is 25° smaller and the second is 25° larger than the third,   (b) one angle is 20° smaller and the second is 50° larger than the third, (c) two of the angles are equal and the third is 48° larger than these.

23 A swimming bath is 10 m longer than it is wide. A swimmer completes three lengths and two widths covering a total distance of 105 m.
    (a) Call the width of the bath x metres and write an expression for the length.   (b) Find, in terms of x the total distance swum. (c) Form an equation in x and solve it.   (d) Deduce the dimensions of the bath.

24 The expression $(2n + 1)$, where n is an integer, gives odd numbers.
    (a) Test this by putting n equal to 9, 19 and 79.   (b) Write down expressions for (1) the next odd number, (2) the previous odd number, (3) the sum of three consecutive odd numbers. (c) Form an equation if the sum of three consecutive odd numbers is 273 and solve it. Hence find the three numbers.

(d) Explain why the sum of three consecutive odd numbers is a multiple of three.

25 A rectangle is 6 cm longer than it is wide. Find its area if its perimeter is 68 cm.

26 On a 72 km journey a driver travels at 40 km/h where there is a speed limit and 60 km/h otherwise. Find the distance over which there is a speed limit if the journey takes 1 hour 20 minutes.

Solve the following pairs of simultaneous equations:

27 $\quad x + y = 12$ $\qquad$ 28 $\quad 2p + q = 16$ $\qquad$ 29 $\quad r + 3s = 1$
$\qquad 2x - y = 18$ $\qquad\qquad 5p + q = 34$ $\qquad\qquad r - 2s = 11$

30 $\quad 2x - 3y = 15$ $\quad$ 31 $\quad 3v - 2w = 13$ $\quad$ 32 $\quad 3y + 2z = 13$
$\qquad 3x + y = 6$ $\qquad\qquad v + 2w = 31$ $\qquad\qquad 2y - 5z = 34$

33 $\quad 3x + y = 9$ $\quad$ 34 $\quad 3a - b = 2$ $\quad$ 35 $\quad 2k + l = 5$
$\qquad 2x + 4y = 11$ $\qquad\qquad 4a - b = 8$ $\qquad\qquad 3k - 2l = 25$

36 $\quad x + 2y = 5x - y = 11$ $\quad$ 37 $\quad 5x - y = y - 3x = 2.$

38 It is given that $y = ax + b$ and that (1) $y = 11$ when $x = 2$ and (2) $y = 17$ when $x = 4$. Form a pair of simultaneous equations in a and b and solve these. Hence find y when $x = 8$.

39 $y = ax^2 + b$ and (1) $y = 17$ when $x = 2$, (2) $y = 42$ when $x = 3$. Find the values of a and b and the value of y when $x = 4$.

40 $y = ax + \dfrac{b}{x}$ and (1) $y = 32$ when $x = 5$, (2) $y = 34$ when $x = 6$. Find the values of a and b and the value of y when $x = 10$.

41 For what values of x and y does $y = 3x + 2$ and $y = 7x - 2$ simultaneously?

42 For what values of p and q does $p = 2q - 5$ and $3p - q = 12$ simultaneously?

43 A dartboard and three sets of darts cost £5·30. Two dartboards and five sets of darts cost £9·90. Find the separate cost of a dartboard and a set of darts.

44 51p buys three ice lollies and four tubs of ice cream. 43p buys five lollies and two tubs. Find the cost of (a) a lolly and (b) a tub.

45 Lunch for three and coffee for two costs £2·55. Lunch for five and coffee for four costs £4·30. Find the cost of (a) lunch and (b) coffee.

Solve the following quadratic equations:

46 $x^2 + 3x + 2 = 0$ 47 $x^2 - 5x + 4 = 0.$

48 $x^2 - 6x + 8 = 0.$ 49 $x^2 + 7x + 12 = 0.$

50 $x^2 + 2x - 15 = 0.$ 51 $x^2 - 7x - 18 = 0.$

52 $2x^2 - x - 3 = 0.$ 53 $3x^2 - x - 10 = 0.$

54 $2x^2 + 11x + 12 = 0.$ 55 $5x^2 + 13x - 6 = 0.$

56 $6x^2 - x - 2 = 0.$ 57 $4x^2 - x - 3 = 0.$

58 $x^2 + x = 0.$ 59 $5x^2 + 3x = 0.$

60 $4x^2 - 6x = 0.$ 61 $3x^2 - 9x = 0.$

62 $x^2 - 121 = 0.$ 63 $x^2 - 64 = 0.$

64 Solve: (a) $x^2 - 5x + 4 = 0$ and $x^4 - 5x^2 + 4 = 0$,
(b) $4s^2 - 5s + 1 = 0$ and $4s^4 - 5s^2 + 1 = 0$,
(c) $p^2 - p - 6 = 0$ and $p - \sqrt{p} - 6 = 0$,
(d) $q^2 + 3q - 10 = 0$ and $q - 3\sqrt{q} - 10 = 0$,
(e) $k^2 - 5k + 6 = 0$ and $1/k^2 - 5/k + 6 = 0$,
(f) $6t^2 + 7t - 3 = 0$ and $6/t^2 + 7/t - 3 = 0$.

65 Form quadratic equations with solutions (a) $x = 5$ or
-2, (b) $y = -3$ or 4, (c) $z = 2\frac{1}{2}$ or 4, (d) $r = \frac{1}{3}$
or 3, (e) $s = 0$ or 6, (f) $t = 3\frac{1}{2}$ or $-3\frac{1}{2}$.

66 The sum 'S' of the first n natural numbers is given by
$S = \frac{1}{2}n(n + 1)$. e.g. $1 + 2 + 3 \ldots + 10 = \frac{1}{2}(10)(10 + 1)$
$= 55$. (a) Calculate S when $n = 12, 18, 30$ and 100. (b) Calculate n when $S = 21, 120$ and 210.

67 The number of diagonals 'd' that can be drawn in a polygon
with n sides is given by $d = \frac{1}{2}n(n - 3)$. e.g. for a hexagon,
$n = 6$ and $d = \frac{1}{2}(6)(6 - 3) = 9$. (a) Calculate d when $n = 4$,
5, 7 and 12. (b) Calculate n when $d = 20, 35$ and 104. (c) Is
it possible to have a polygon with 40 diagonals? Justify your
answer.

68 A rectangle is 6 cm longer than it is wide. Its area is
216 cm^2. Find (a) its width, (b) its perimeter.

69 A triangle with a base of length 'b' has a height 3 cm less
than this. Calculate b if the area of the triangle is 350 cm^2.

70 A piece of wire 72 cm long is bent to form a rectangle of
area 320 cm^2. Calculate the lengths of the sides of the rectangle.

71 One side of a right angled triangle is 3 cm longer than a
second side. The hypotenuse is 3 cm longer again. Calculate the
lengths of the sides of the triangle.

72   A rectangle has sides x and $x + 4$. Calculate   (a) its area when $x = 10$, 12 and 25 cm,   (b) its sides when its perimeter equals 88, 104 and 272 cm,   (c) its sides when its area is 60, 165 and 285 cm^2.

The Use of Formulae

Exercise 16

1   Write formulae for   (a) the number of days D in W weeks (b) the number of seconds S in H hours,   (c) the number of rotations R in D degrees,   (d) the number of legs L in P pyjamas,   (e) the number of buttons B on P pyjamas with C on each coat and T on each pair of trousers,   (f) the number of fingers F on G pairs of gloves,   (g) the number N of square tiles of side S to cover a floor of length L and width W,   (h) the total time for a journey of K kilometres at u km/h for the first third and v km/h for the remainder,   (i) the length L of a ladder with R rungs a distance D apart and a space S at both ends,   (j) the number of traffic wardens W for M meters at T meters per warden,   (k) the cost C of parking for H hours at P pence for the first hour and $2P$ pence for subsequent hours.

2   A cube has a side of length S. Write down formulae for (a) its volume,   (b) its total surface area,   (c) the total length of its edges,   (d) the length of one of its longest diagonals.

3   A car travels at U km/h for M minutes and V km/h for N minutes. Write down formulae for   (a) the separate distances travelled,   (b) the total distance travelled,   (c) the average speed.

4   A crate contains B boxes and there are P pairs of stockings in each box. Find expressions for   (a) the number of boxes (1) in C crates (2) for N pairs of stockings,   (b) the number of stockings in (1) B boxes (2) C crates,   (c) the number of crates for (1) X boxes (2) S pairs of stockings.

5   The formula for the volume V of a sphere of radius r is $V = \frac{4}{3}\pi r^3$. Write down expressions for   (a) the volume of N spheres,   (b) the volume of H hemispheres,   (c) the number of spheres that can be made from a cube of wax of side S.

6   Write down formulae for the number of posts P placed a distance D apart,   (a) along a line of length L,   (b) around a

square of side S, (c) around a rectangle of sides R and Q, (d) around a circle of radius R.

7 $S = (2n - 4)$ right angles, gives the sum S of the interior angles of a polygon with 'n' sides. Calculate S when n equals 6, 8, 10 and 24.

Rearrange the formula to make n the subject and calculate n when S equals 10, 24, 36 and 50 right angles. What restriction is there on the value of S?

8 The area A of a trapezium with parallel sides of lengths 'a' and 'b' a distance 'h' apart is given by $A = \frac{1}{2}(a + b)h$. Rearrange this formula to make (1) h, (2) a and (3), b the subject.

Calculate the missing element if (1) $a = 6$, $b = 10$ and $h = 9$, (2) $A = 144$, $a = 8$ and $b = 10$, (3) $A = 180$, $a = 12$ and $h = 10$, (4) $A = 240$, $b = 16$ and $h = 15$.

9 Transform the following formulae to the subject indicated.

Formula	$C = 2\pi r$	$V = lbh$	$I = \dfrac{prt}{100}$	$V = \frac{1}{3}Ah$
Subject	r	l,b,h	p,r,t	A,h

Formula	$A = \pi rl$	$S = \frac{1}{2}(a + l)n$	$S = \frac{1}{2}(u + v)t$
Subject	r,l	n,a,l	t,u,v

10 Change the subject of the following formulae to the symbol indicated.

Formula	$A = \pi r^2$	$A = 4\pi r^2$	$E = mc^2$	$V = \dfrac{4}{3}\pi r^3$
Subject	r	r	m,c	r

Formula	$r = \sqrt{\dfrac{3V}{\pi h}}$	$t = \sqrt{\dfrac{2s}{a}}$
Subject	h,V	a,s

11 Rearrange the following formulae to change the subject as indicated

Formula	$\dfrac{1}{v} + \dfrac{1}{u} = \dfrac{1}{f}$	$\dfrac{1}{v} - \dfrac{1}{u} = \dfrac{1}{f}$	$\dfrac{1}{R} = \dfrac{1}{R_1} + \dfrac{1}{R_2}$	$\dfrac{1}{a} = \dfrac{2}{b} + \dfrac{3}{c}$
Subject	f,u	f,v	R,R_1,R_2	a,b,c

12 Express x in terms of y if (a) $y = 4x + 5$,
(b) $y = 3x - 2$, (c) $y = \frac{1}{3}(4x + 2)$, (d) $y = \frac{3}{4}(3x - 8)$,
(e) $y = 2x^2 + 3$, (f) $y = 5x^2 - 4$, (g) $y = \frac{1}{4}(3x^2 - 5)$,
(h) $y = \sqrt{6x + 7}$, (i) $y = \sqrt{3x^2 + 4}$, (j) $y = \sqrt{2x^2 - 3}$.

13   Transform the following to express the second letter in terms of the first.

(a) $y = \dfrac{4x + 2}{3x}$,   (b) $R = \dfrac{K + 7}{K - 7}$,   (c) $s = \dfrac{3t + 4}{2t - 5}$,

(d) $p = \dfrac{5 - q}{5 + q}$,   (e) $M = \dfrac{5N}{5 - N}$

14   Given $R = (2s + 3)$ and $s = \frac{1}{2}(t - 5)$, express   (a) s in terms of R,   (b) t in terms of s,   (c) R in terms of t.

Relations

Exercise 17

1   Write out in full the product set PQ, where $P = \{3, 6, 9\}$ and $Q = \{a, b, c, d\}$. How does the product set QP differ from the one just completed?

2   $R = \{1, 2, 3, 4\}$ and $S = \{x, y, z\}$. How many elements are there in the product set RS? List those elements of RS that are of the form   (a) $(3, —)$,   (b) $(—, x)$. Of what product set is the element $(y, 4)$ a member?

3   Take the sets $M = \{m, u, g\}$ and $N = \{1, 3, 5\}$ and write out the cartesian product set MN. How many ordered pairs are there in this product set?

How many ordered pairs are there in MN with   (a) first element u,   (b) second element 5,   (c) first element m and second element 3,   (d) first element 1 and second element g?

Write down the inverses of $(m, 1)$ and $(g, 5)$, and write out the product set NM.

4   $T = \{t, o, n\}$ and $U = \{u, p\}$. Tabulate the cartesian product set TU. How many ordered pairs are there in TU with (a) first element t,   (b) second element p,   (c) first element a vowel,   (d) either element a vowel?

Tabulate the inverse product set UT. How many possible relations can be formed from the ordered pairs of UT? How many from those of   (1) TT,   (2) UU?

5   Form the product set PQ where $P = \{2, 3, 5\}$ and $Q = \{4, 6, 12, 15\}$. Is the product set QP the same as PQ?

Say if the following ordered pairs are elements of PQ or QP $(3, 6)$, $(4, 2)$, $(6, 5)$, $(5, 15)$, $(3, 3)$.

6   In question 5, how many elements of PQ are there in the relation,   (a) .. is less than ..,   (b) .. leaves a remainder 1

when divided into . ., (c) . . has a factor in common with . ., (d) {x, y: y = 3x}, (e) . . is not a factor of . .

7 Three athletes, Will (W), Phil (P) and Jill (J) train on various nights of the week as follows: (P, Mon), (P, Thur), (W, Tue), (W, Wed), (J, Sat), (J, Sun). Illustrate this relation by a diagram, a table and a graph.

Is this a many-many relation? Why is it not a mapping?

8 Draw graphs to illustrate the following relations.
  (a) (2, 3) (2, 5) (3, 1) (4, 6), (b) (1, 1) (2, 3) (3, 5) (4, 7), (c) (1, 4) (2, 4) (1, 6) (2, 7), (d) (2, 4) (−3, 9) (−4, 16) (5, 25).

Classify the relations as many-many etc. Say which relations are functions. Express the functions in the form $f : x \rightarrow \ldots$

9 Repeat question 8 for
  (a) (2, 4) (3, 4) (3, 1) (3, 2) (b) (1, 1) (−1, 1) (2, $\frac{1}{4}$) (−2, $\frac{1}{4}$) (c) (−1, −1), (1, 1) (2, 8) (3, 27) (d) (3, 4) (3, 5) (2, 3) (5, 6).

10 Sampson (S), Goliath (G) and Strong John (N) are members of a Physical Fitness Club that offers classes in Weight-Lifting (W), Judo (J), Boxing (B) and Wrestling (R).

Illustrate and explain the relation (S, B) (S, W) (G, W) (G, B) (G, R) (N, B) (N, R).

(a) Who takes part in most activities? (b) Which class has least support? (c) What type of relation is this? (d) Explain why this is not a mapping.

11 Illustrate the following relations and say whether they are many-many etc.
  (a) (a, v) (m, c) (t, c) (e, v) (b) (4, 7) (5, 8) (1, 4) (2, 5) (c) (p, q) (a, b) (u, v) (y, z) (d) (1, 1) (−1, 1) (2, 16) (−2, 16).

Which of the above are functions? Express each relation in words.

12 Given $f : x \rightarrow 3x^2 - 3$, (a) find the images of $f(4)$ $f(6)$ $f(-6)$ $f(\frac{1}{2})$ and $f(-\frac{1}{2})$, (b) say whether this function is one-one or many-one, (c) state the range if the domain of x is $-2 < x < 2$, (d) is the inverse relation a function?

13 Repeat question 12 for (1) $f : x \rightarrow 5x - 2$, (2) $f : x \rightarrow \dfrac{60}{x + 2}$.

14 Given $f : x \rightarrow 3x + 2$ and $g : x \rightarrow x^2$, (a) find $f(3)$ $f(7)$ $g(4)$ and $g(-4)$, (b) say whether these functions are one-one or many-one, (c) find x if (i) $f(x) = 14$, (ii) $g(x) = 36$, (d) express in simplest form $f[g(x)]$ and $g[f(x)]$. Are the inverse relations functions?

15 Given $f:x \to 2x + 5$, $g:x \to 2x^2$ and $h:x \to \dfrac{1}{x}$ find
(a) $f(3)$ $g(4)$ $h(5)$ $f(-4)$ $g(\frac{1}{2})$ $h(0\cdot2)$, (b) $f[g(2)]$;
$g[f(2)]$; $g[h(5)]$; $h[g(5)]$, (c) $f\{g[h(x)]\}$; $f\{h[g(x)]\}$.

16 Complete the following table:

Domain	Integers	Whole numbers		Angles 0–90°
Function	$f:x \to 2x^2$		$f:x \to 3x+1$	$f:x \to \sin x$
Range		Multiples of 5.	1, 7, 10 . . .	

Domain	Whole numbers	Angles 0–90°	
Function	$f:x \to 2x+1$	$f:x \to \tan x$	$f:x \to x^2$
Range			non-negative integers.

17 Find, where possible, the inverses of the following functions. (a) $f:x \to 3x + 2$, (b) $f:x \to 2x - 3$,
(c) $f:x \to x^2$, (d) $f:x \to \frac{1}{3}(x + 8)$, (e) $f:x \to x^3 + 4$.

18 Write down the inverse of each of the following relations.
(a) . . . is the brother of . . . for pupils in a boys school.
(b) . . . is the sister of . . . for pupils in a mixed school.
(c) . . . has more chocolate on than . . . for éclairs in a box.
(d) . . . is the mother of . . . for a set of females.
(e) . . . is twice as long as . . . for a set of snakes.
(f) . . . is 50% longer than . . . for a set of ladders.
(g) . . . has a factor in common with . . . for a set of numbers.
   What is the special feature of relations 'a' and 'g'?

Graphical Work

Exercise 18

1 Take O as the origin and plot the points K(6, 8) and L(15, 8). Find (a) the lengths KL, OK and OL, (b) the area OKL, (c) the co-ordinates of P if figure OKPL is symmetrical about KL.

2 In the previous question find the co-ordinates of the fourth vertex of the parallelogram with (a) OL, (b) OK, (c) KL, as a diagonal.

3 Call the origin O and plot the points P(0, 2), Q(5, 14) and R(5, 2). Find (a) the length PQ, (b) the areas of triangles PQR and OPQ, (c) the equations of the sides of triangle PQR.

4   In question 3 find the value of
(1) a if PQ = QS with S(a, 2),
(2) b if PQ = QT with T(0, b),
(3) c if PQ = QU with U(c, 14).

5   Plot the points P(4, 4), Q(7, 8) and R(11, 4). Find   (a) the lengths PR and PQ,   (b) the area of triangle PQR,   (c) the co-ordinates of the fourth vertex of the parallelogram with (1) QR, (2) PR and (3) PQ as a diagonal,   (d) the co-ordinates of T if PQRT is a kite.

6   Given the straight lines   (1) $y = 2x + 3$,   (2) $y = 5x + 4$, (3) $y = 3x - 1$,   (4) $y = 4 - 2x$,   (5) $x + y = 7$ and   (6) $3x + 2y = 12$;   find   (a) their gradients,   (b) the coordinates of the points where the graphs cut the x and y axes,   (c) the areas of the triangles between each line and the two axes.

7   Sketch the six lines in question 6. Find the six pairs of values of p and q if (4, p) and (q, 2) lie on the lines.

8   Using a scale of 1 cm to 1 unit for both axes plot the points tabled below and join to form two straight lines.

x	−2	0	2
y	−1	3	7

x	−1	0	2
y	9	7	3

Find   (a) the co-ordinates of the point of intersection of the two lines,   (b) the area between the two lines and the x-axis, (c) the area between the two lines and both the axes.

9   Find the equations of the two lines in question 8.

10   Consider the following statements about the graph of $y = 3x - 2$ and say if they are true or false.
(a) The gradient of the graph is $+3$.   (b) The point (5, 12) does not lie on the line.   (c) The graph cuts the x-axis at ($\frac{2}{3}$, 0). (d) The area between the graph and the two axes is $\frac{1}{3}$ square units.   (e) The points (0, 0) and (2, 3) are on opposite sides of the line.

11   Solve the following simultaneous equations by graphical means:
(a) $y = 2x - 5$ (b) $y = 2x - 1$ (c) $y = x + 5$
    $y = 4 - x$ $x + y = 5$ $4x + y = 0$.

12   A car hire firm offers two methods of payment:
Tariff A. A basic charge of £2 plus 1p per kilometre.
Tariff B. No basic charge but $2\frac{1}{2}$p per kilometre.

54

Plot graphs to represent these tariffs using a scale of 1 cm to 10 km horizontally and 2 cm to £1 vertically. From your graphs estimate: (a) the two charges for 80 km and 180 km, (b) the distance travelled for £3·50 on the two schemes, (c) which method is cheaper and by how much for 100 km and 140 km, (d) the distance for which the two costs are equal.

13 Plot the letters shown in Fig. 18.1. Write down the equations of the lines forming the letters.

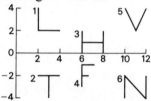

Fig. 18.1

14 In Fig. 18.2, find (a) the co-ordinates of the points P, Q, R and S, (b) the length QR, (c) the area OPQ, (d) the area PQRS.

Fig. 18.2

Fig. 18.3

15 Fig. 18.3 illustrates the graph of $y = (1 + x)(5 - x)$. Find, (a) the co-ordinates of P, Q, R and the maximum S, (b) whether the points $(2, 9)$, $(-2, -7)$, $(7, -16)$ lie on the curve, (c) the solutions of $x^2 - 4x - 5 = 0$.

16 Use the table below to plot a graph of $y = 2x^2 - x - 3$. Take 2 cm to one unit for x and 1 cm to one unit for y.

$x =$	-2	-1	0	1	2	3
$y =$	7	0	-3	-2	3	12

Use your graph and a suitable straight line to solve
(a) $2x^2 - x - 3 = 0$, (b) $2x^2 - x - 3 = 5$,
(c) $2x^2 - x - 5 = 0$, (d) $2x^2 - x = 0$,
(e) $2x^2 - 2x - 3 = 0$, (f) $2x^2 - 2x - 5 = 0$.

17 Plot a graph of $y = x^2$ for values of x from -4 to $+4$ using a scale of 2 cm to one unit for x and 1 cm to one unit for y. Use your graph to (a) find $1·7^2$, $2·3^2$, $3·3^2$, (b) find $\sqrt{8}$, $\sqrt{10}$, $\sqrt{15}$, (c) solve $x^2 = x + 6$, $x^2 = 3x + 2$, $x^2 + 2x - 5 = 0$.

18   Plot a graph of $y = 4x - x^2$ for values of x from -1 to $+5$. Use your graph to find   (a) solutions of $4x - x^2 = 0$, $4x - x^2 = 2$, $x^2 - 4x + 3 = 0$,   (b) the gradients of the graph at $x = 1$, 2 and 4.

Explain how the solutions of $x^2 - 3x = 0$ can be found by drawing the line $y = x$.

19   Use the graph in question 18 to find the maximum value of $4x - x^2$. Without drawing any further graphs find   (a) the maximum values of $2(4x - x^2)$ and $5 + 4x - x^2$,   (b) the minimum value of $x^2 - 4x$.

20   A cannonball is fired from the top of a cliff. Its height h metres above the cliff edge t seconds after firing is given by $h = 40t - 10t^2$. Calculate values of h for $t = 0$, 1, 2, 3, 4, and 5 seconds.

Plot a distance–time graph for the flight of the cannonball using a scale of 4 cm to one second horizontally and 1 cm to 10 metres vertically.

21   Use the graph from question 20 to estimate   (a) the height of the ball at 3·3 seconds,   (b) the two times when the ball was 25 m high,   (c) the time for which the ball was above 35 m, (d) whether the ball was moving upwards or downwards at 4·5 seconds and whether it was above or below the top of the cliff at that time.

22   A skier slides down a ski run and the distance he slides s metres in a time t seconds is given by $s = \dfrac{5}{2} t^2$. Make a table of values of s for $t = 0$, 1, 2, 3, 4, 5 and 6 seconds.

Plot a graph of distance against time using a scale of 2 cm to one second horizontally and 1 cm to 5 metres vertically. Use your graph to estimate   (a) the distances travelled in 1·5, 2·5 and 3·5 seconds,   (b) the time to cover the first, second and third 30 metres,   (c) the speed of the skier at 3 seconds.

23   An aircraft takes off vertically. Its height in metres at intervals of 5 seconds is given by

Time(s)	0	5	10	15	20	25	30	35
Height(m)	0	35	110	225	370	520	660	750

Plot a graph of height against time using a scale of 2 cm to 5 s horizontally and 1 cm to 50 m vertically. Use your graph to estimate   (a) the heights at 12, 22 and 32 seconds,   (b) the time to clear 250 m,   (c) the average rate of climb,   (d) the maximum rate of climb.

24   A club organises a charity walk which starts at 10 00 h from the village of Lower Lypp and covers the 20 km to the

village of Upper Lypp. Members walk at 8 km/h for 45 minutes and then rest for 15 minutes. They follow this pattern until 11 45 h when they take a lunch break of 1 hour, restarting their walking pattern at 12 45 h.

Plot a distance–time graph for the walk using a scale of 4 cm to 1 hour horizontally and 1 cm to 1 km vertically. Find (a) how far from Upper Lypp they take lunch, (b) at what time they complete the walk, (c) the average speed of the walkers.

25 A car carrying hot soup leaves Lower Lypp (see question 24) at 11 00 h and drives to be at the lunch point 15 minutes before the walkers. Find (a) the average speed of the car, (b) when and where the car passed the walkers.

26 At 14 00 h a cyclist sets out from Uphill to cycle the 30 km to Downdale. He rides at 15 km/h for one hour when he has trouble with his front brake. He spends 10 minutes trying to repair this but gives up and spends 5 minutes telephoning his relatives at Downdale. They say they will come for him in a car starting at 15 45 h. Meanwhile he pushes his cycle at 4 km/h towards Downdale. If the car travels at 35 km/h, find, by drawing a distance–time graph (a) when and where the car met the cyclist, (b) how far the cyclist walked.

Binary Operations and Groups

Exercise 19

1 $a \circ b = \dfrac{a + b}{2}$, e.g. $43 \circ 93 = \dfrac{43 + 93}{2} = 68$.

(a) Calculate 74 o 126, 145 o 23, 33 o 192.
(b) Find x if $x \circ 6 = 14$. (c) Simplify $x \circ 5x$. (d) Does $a \circ b = b \circ a$? (e) Does $(a \circ b) \circ c = a \circ (b \circ c)$?

2 $x \circ y = x^2 + y^2$.
(a) Calculate 3 o 4, 4 o 5, 2 o (3 o 4), (2 o 3) o 4.
(b) Does $x \circ (y \circ z) = (x \circ y) \circ z$? (c) Does $x \circ y = y \circ z$?
(d) If $a \circ b = a \circ c$, what can be said about b and c?

3 $p \circ q = \dfrac{p + q}{pq}$.
(a) Calculate 2 o 3, 3 o 4, $\frac{1}{2} \circ \frac{1}{4}$. (b) Does $p \circ q = q \circ p$? (c) Does $(p \circ q) \circ r = p \circ (q \circ r)$? (d) Find p if $p \circ p = \frac{1}{3}$. (e) Is it possible to have $p \circ q = 1$? Give an example.

4 Say whether the following sets are closed under the given operation.
(a) {even numbers}, under addition. (b) {odd numbers}, under addition. (c) {integers}, under subtraction. (d) {whole numbers}, under division. (e) {prime numbers}, under multiplication. (f) {multiples of 6}, under addition.

5 Which of the following sets are closed under the given operation?
(a) $\{-1, 0, 1\}$, multiplication. (b) $\{-1, 0, 1\}$, addition. (c) $\{-2, -1, 0, 1, 2\}$, multiplication. (d) {odd numbers}, squaring. (e) {square numbers}, cubing.

(f) $\left\{ \begin{pmatrix} 1 & 0 \\ 0 & 1 \end{pmatrix} \quad \begin{pmatrix} -1 & 0 \\ 0 & -1 \end{pmatrix} \right\}$, matrix multiplication.

6 Name the identity element in the following sets under the given operation.
(a) {whole numbers}, addition. (b) {integers}, subtraction. (c) {integers}, multiplication. (d) {2 × 2 matrices}, addition. (e) {2 × 2 matrices}, multiplication.

7 State the inverse of the following elements under the operation given.
(a) 6, addition of integers. (b) 12, multiplication of real numbers. (c) $\frac{5}{8}$, multiplication of rational numbers.

(d) $\begin{pmatrix} 4 \\ -2 \end{pmatrix}$, vector addition. (e) $\begin{pmatrix} 2 & 3 \\ 1 & 4 \end{pmatrix}$, matrix addition.

(f) $\begin{pmatrix} 3 & 1 \\ 5 & 2 \end{pmatrix}$, matrix multiplication. (g) Letter W, reflexion in the x-axis.

8 Illustrate by means of Venn diagrams
(a) $P \cap Q = Q \cap P$, (b) $(P \cap Q) \cap R = P \cap (Q \cap R)$,
(c) $(P \cup Q) \cup R = P \cup (Q \cup R)$.

9 $P - Q$, is defined as the elements of set P that are not elements of set Q. (The shaded region in Fig. 19.1.)
(a) Does $P - Q = Q - P$?
(b) Does $(P - Q) - R = P - (Q - R)$?
(c) Does $P - (Q - R) = (P - Q) - (P - R)$?

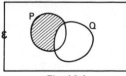

Fig. 19.1

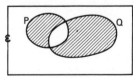

Fig. 19.2

10 $P \triangle Q$, is the set of elements in P and not in Q or in Q and not in P. (The shaded region in Fig. 19.2.)

(a) Does $P \triangle Q = Q \triangle P$?
(b) Does $(P \triangle Q) \triangle R = P \triangle (Q \triangle R)$?

11 State four conditions to be satisfied if the set 'S' and the operation 'o' is to form a group.
Is it necessary for the operation to be commutative?

12 Copy and complete the multiplication table opposite for the numbers 1, 3, 5 in arithmetic modulo 6.

$\times$	1	3	5
1			
3		3	
5			1

Is the table closed? Is there an identity element? Is the operation associative? Why is this not a group?

13 Using the numbers 1, 3, 5 and 7 compile a multiplication table for arithmetic modulo 8. Show that this is a group.

14 Use your table from question 13 to (a) find the inverse of 7, (b) decide if the group is commutative, (c) simplify 3×5^3, (d) find the solutions of $x^2 \equiv 1$.

15 Make a multiplication table, modulo 10, for the numbers 1, 3, 5, 7 and 9. Is the table closed? Is there an identity element? Explain why this is not a group.

16 Compile a multiplication table, modulo 10, for the numbers 2, 4, 6 and 8.
(a) Is the table closed?
(b) Does $(4 \times 6) \times 8 = 4 \times (6 \times 8)$?
(c) What is the identity element? (d) Give the inverses of 4 and 8. (e) Which elements are self inverse? (f) Is this a group? (g) Is this a commutative group?

17 $s \sim t$, is defined as the numerical difference between s and t, e.g. $8 \sim 5 = 3$ and $7 \sim 12 = 5$.
Copy and complete the table opposite.
Say whether the following features are present or not.
(a) closure, (b) associativity,

$\sim$	0	4	8	12
0		4		
4				
8				4
12				

(c) an identity element, (d) inverse elements. (e) Is this an example of a group?

18 $R -$ is for 'Right turn', $L -$ is for 'Left turn', $A -$ is for 'About turn' and $S -$ is for 'Stay as you are'.
(a) Compile a table for $\{R, L, A, S\}$ and the operation 'followed by'. (b) Identify the identity element. (c) What is the inverse of R? (d) Which command is self inverse? (e) Do the above form a group? (f) Do the above form a commutative group?

19 Refer to the rectangle opposite.

1
2

Operator S_1 means shade the upper half
of the rectangle. Operator S_2 means shade the lower half of the
rectangle. S_3 means shade all the rectangle. R means rub out all
shading.
(a) Compile a table for $\{S_1, S_2, S_3, R\}$ and the operation
'followed by'. (b) Is the system closed? (c) Is there an
identity element? (d) Give two reasons why this is not a group.

20 Three soldiers stand in a row in positions P_1, P_2 and P_3.
Operator M_1 means move the soldier in position P_1 to the other
end of the row. M_2 means move the soldier in P_3 to the other
end. M_3 means swop the soldiers in P_1 and P_2. M_4 means swop
those in P_2 and P_3. M_5 means swop those in P_1 and P_3. N means
'do nothing'.
(a) Compile a table for $\{M_1, M_2, M_3, M_4, M_5, N\}$ and the
operation 'followed by'. (b) Show that these form a group.

21 Show that the set of rotations through 120°, 240° and 360°
together with the operation 'followed by', forms a group.

22 Show that the set of rotations 60°, 120°, 180°, 240°, 300°
and 360° with the operation 'followed by', forms a group.
What is the relation between the group in question 21 and
this group?

23 Show that $\{\{2, 4\}, \{2\}, \{4\}, \phi\}$ and the operation $\cap$, do not
form a group.

24 Show that the matrices
$$\begin{pmatrix} 1 & 0 \\ 0 & 1 \end{pmatrix}, \begin{pmatrix} 1 & 0 \\ 0 & -1 \end{pmatrix}, \begin{pmatrix} -1 & 0 \\ 0 & -1 \end{pmatrix} \text{ and } \begin{pmatrix} -1 & 0 \\ 0 & 1 \end{pmatrix}$$
form a group under matrix multiplication.

25 Figure 19.3 shows a sort of 'purse' that has a
zip fastener at the top and the bottom.
Operator Z_1 means reverse the position of the upper zip. Z_2
means reverse the lower zip. B means reverse both zips. N
means 'do nothing'.
(a) Compile a table for $\{Z_1, Z_2, B, N\}$ and the operation
'followed by'. (b) Is the table closed? (c) What is the identity
element? (d) What can be said about the elements and their
inverses? (e) Does the system form a group?

26 Which of the following are group tables?

(a)

+	O	E
O	E	O
E	O	E

(b)

~	0	1	2
0	0	1	2
1	1	0	1
2	2	1	0

(c)

+	0	1
0	0	1
1	1	2

(d)

+	R_1	R_2
R_1	R_2	R_1
R_2	R_1	R_2

(e)

×	0	1
0	0	0
1	0	1

Explain why the others are not group tables.

27   Compile and compare tables for   (a) Addition modulo 3 with the elements 0, 1, 2.   (b) Rotations $R_1 = 120°$, $R_2 = 240°$ and $R_3 = 360°$ with the operation 'followed by'.

Are these groups? Do they have the same form i.e. are they isomorphic?

28   Repeat question 27 for   (a) addition modulo 4 with elements 0, 1, 2, 3 and   (b) rotations of 90°, 180°, 270°, 360° and the operation 'followed by'.

29   Make a table for multiplication modulo 5 using the elements 1, 2, 3, 4. Is this a group? Is this isomorphic with the tables in question 28? If so, name the corresponding elements.

30   Compare the table from question 29 with that for the commands 'Right turn', 'Left turn', 'About turn' and 'Stay as you are', together with the operation 'followed by'. Is there a common form? State any correspondence between the elements.

31   These questions concern a sack, which is open at one end. Operation T stands for 'Turn the sack inside out'. L stands for 'Leave the sack as it is'.

(a) Compile a table for T, L and the operation 'followed by'. (b) Is this a group?   (c) Compare your table with that for the addition of odd and even numbers. Is there an isomorphism?

32   Say which of the following tables are isomorphic.

(a)

o	p	q
p	p	q
q	q	p

(b)

0	s	t
s	s	s
t	t	t

(c)

+	O	E
O	E	O
E	O	E

(d)

×	O	E
O	O	E
E	E	E

(e)

+	0	1
0	0	1
1	1	0

(f)

×	0	1
0	0	0
1	0	1.

Topology

Exercise 20

1   Are the following topologically equivalent?
 (a) A carrot, a piece of sliced carrot, a piece of diced carrot.
(b) A pineapple, a pineapple chunk, a pineapple ring, a pine-apple piece.   (c) a serviette ring, a key ring, a wedding ring.

2   Topologically, which is the odd one out in the following?
(a) A cough drop, a toffee apple, a peanut, a mint-with-a-hole, a humbug.   (b) An American doughnut, an inflated inner tube, a British doughnut, a lifebelt.   (c) A gramophone record with a single hole, a mug with one handle, a flower pot with one drain hole, a tumbler.

3   Partition the following items of jewellery into topologically equivalent sets: a cut diamond, a bracelet, a single bead, one drop pearl, a signet ring.

4   Explain the topological difference, if any, between   (a) a spaghetti hoop and a spaghetti string,   (b) a 50p piece and a 2p piece,   (c) a pullover and a cardigan – ignore buttonholes.

5   Partition the letter networks in Fig. 20.1 into topologically equivalent sets.

Fig. 20.1

6   (a) Name six fruits that are topologically equivalent to a sphere.
(b) Can you think of any fruits equivalent to a torus?

7   Which pairs of networks in Fig. 20.2 are equivalent?

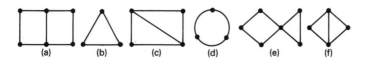

Fig. 20.2

62

8 Arrange the networks in Fig. 20.3 into topologically equivalent sets.

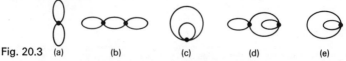

Fig. 20.3 (a)   (b)   (c)   (d)   (e)

9 The network shown in Fig. 20.4 is drawn on a balloon and the balloon is then inflated. Which of the following remains unchanged under this transformation?

(a) The length PQ, (b) RST is a 3-sided figure, (c) the area enclosed by RST, (d) Q is between P and R, (e) angle QRT, (f) X is inside RST.

Fig. 20.4   Fig. 20.5

10 A gardener scratches Fig. 20.5 on a young vegetable marrow, which grows and grows. Say which of the following properties of the figure are invariant.

(a) The number of arcs, (b) the length of the arcs, (c) the three arcs meet in one point, (d) the X's are inside the arcs, (e) the figure divides the surface of the marrow into 4 regions.

11 Is Figure 20.6 a simple closed curve?

State whether the points X, Y and Z are inside or outside the curve.

Fig. 20.6   Fig. 20.7

12 Repeat question 11 for Fig. 20.7.

13 Refer to the parts of Fig. 20.8 for the following questions. 1. How many inside regions does each figure possess? 2. Which figure is a simple closed curve? 3. In which figures is it possible to join X to Y without crossing any line of the figure?

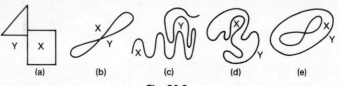

(a)   (b)   (c)   (d)   (e)

Fig. 20.8

14 (a) Houses H_1, H_2 and H_3 in Fig. 20.9 are to be joined up to Electricity E and Water W. No electricity or water mains are to cross. Show how this can be done.

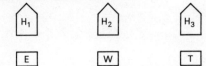

Fig. 20.9

(b) How many of the houses can now be connected to the Telephone Exchange T, still with no lines crossing?

(c) Explain why all three houses cannot be connected to all three services without a pair of lines crossing.

15 Tabulate the number of nodes N, regions R, and arcs A, for the networks in Fig. 20.10. Include the outside region. Calculate $N + R - A$ in each case.

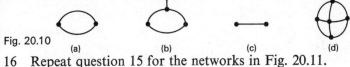

Fig. 20.10   (a)   (b)   (c)   (d)

16 Repeat question 15 for the networks in Fig. 20.11.

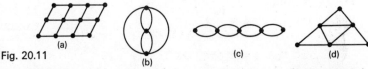

Fig. 20.11   (a)   (b)   (c)   (d)

17 Which of the networks in questions 15 and 16 are unicursive? i.e. Which can be drawn completely without going over any line twice and without taking the pencil off the page?

18 1. Which of the networks in Fig. 20.12 are unicursive?
2. For those that are not, what is the least number of arcs to be added to make them unicursive?

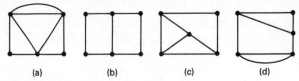

Fig. 20.12   (a)   (b)   (c)   (d)

19 Figure 20.13 shows two 3-node networks.
Why are they given this name?
Count N, R and A for both of these networks. Calculate a – the average number of arcs per region. Calculate 3N, 2A and aR and comment on the answers. Test that $a = 6 - 12/R$.

Fig. 20.13

20 Repeat question 19 for the 3-node network in Fig. 20.14.

Fig. 20.14

21 Draw another 3-node network and test that $3N = 2A = aR$.

22 Figure 20.15 shows two 4-node networks.
(a) Calculate N, R, A and $N + R - A$ in both cases.
(b) Calculate a – the average number of arcs per region.
(c) Compare the values of 4N, 2A and aR.

Fig. 20.15   Fig. 20.16

23 Repeat question 22 for the 4-node network in Fig. 20.16.

24 Draw another 4-node network and test that $4N = 2A = aR$. Can you find the relation between a and R in this case?

25 Figure 20.17 shows the streets around which a postman has to deliver letters. (a) Is it possible to plan the postman's walk

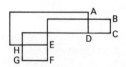

Fig. 20.17

so that he visits each street once only? (b) Has he a choice of starting points? (c) If he must start from A, what is his least amount of double walking?

26 Figure 20.18 shows 4 barges that are connected to each other and to the canal bank by gangplanks.

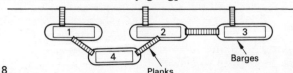

Fig. 20.18   Planks

(a) Represent the situation by a network with nodes for barges and land, and arcs for gangplanks. (b) Is it possible to walk over every gangplank in one continuous tour? (c) Is it possible starting and ending on dry land?
A fifth barge arrives carrying one gangplank. Show in a diagram to which barge it should attach its gangplank so that the

bargee may walk from his barge over every plank and end up on dry land.

27 Calculate V, the number of vertices, F, the number of faces and E the number of edges for (a) a cube (b) a pyramid with a square base, (c) a tetrahedron, (d) a hexagonal pyramid, (e) a hexagonal prism.
    Find the value of V + F − E in each case.

28. (a) Calculate V + F − E for the cuboid in Fig. 20.19.

Fig. 20.19

(b) Make the same calculation for the cuboid with the corner A removed as indicated.
(c) Find V + F − E for the cuboid with all eight corners removed as at A.

29 What is the least number of colours needed to shade the regions in each part of Fig. 20.20, if no two regions with a common boundary are to be the same colour?

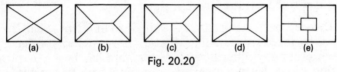

Fig. 20.20

30 Repeat question 29 for the 5 parts of Fig. 20.21.

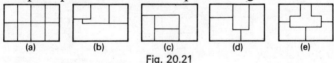

Fig. 20.21

31 Draw Schlegel diagrams of a cube and a triangular prism. Use your diagrams to decide which requires more colours if any two faces with a common edge are to be differently coloured.

32 Draw Schlegel diagrams of (a) a cuboid, (b) a pyramid with a rectangular base, (c) a truncated pyramid with a rectangular base.
    Are any of these alike? What is the minimum number of colours required to shade the faces, as in question 31?

33 Repeat question 32 for (a) a cylinder, (b) a cone, (c) a truncated cone.

34 Draw two different Schlegel diagrams of a triangular prism, (a) removing a triangular base, (b) removing a rectangular face.

Show that both give the same minimum number of colours to paint the faces as above.

35 With the help of a Schlegel diagram, decide the minimum number of colours required to shade the faces of Fig. 20.22.

Can a framework be made in the form of Fig. 20.22, using one piece of wire only?

Fig. 20.22   Fig. 20.23

36 Figure 20.23 shows a Schlegel diagram for a regular octahedron.

(a) Sketch the original octahedron.   (b) Find the least number of colours needed to colour the faces.   (c) What is the minimum number of pieces of wire needed to form the octahedron?

37 What is the minimum number of pieces of wire needed to form   (a) a cube,   (b) a triangular prism,   (c) a tetrahedron, (d) a truncated pyramid with a square base?

38 In each part of Fig. 20.24 the aim is to link 1 to 1, 2 to 2 and 3 to 3, by three separate lines that do not cross each other and do not cross any other lines. One diagram has been completed for you. Find the two other diagrams in which this is possible.

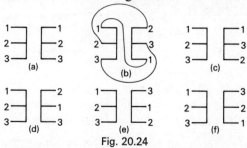

Fig. 20.24

What is the common feature of the three diagrams in which this is not possible?

39 In Fig. 20.25, three skiers P, Q and R set out from the pos-

Fig. 20.25   Fig. 20.26

itions shown to ski to the hut H, which they enter by doors p, q and r respectively. Find suitable paths for them to follow if their ski tracks are not to cross.

40   Repeat question 38 for the four skiers P, Q, R and S in Fig. 20.26.

41   Figure 20.27 shows airports A_1, A_2, A_3, A_4 and A_5 and the air routes between them. The arrows indicate the direction(s) of flight.

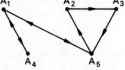

Fig. 20.27

(a) How many landings are made on the journey from (1) A_1 to A_3, (2) A_2 to A_4?   (b) State the orders of airports to touch ground at each once only.   (c) Which additional route is preferable A_4 to A_2 or A_3 to A_2?   Give reasons.

42   Figure 20.28 shows a system of one and two way streets in a busy market town.

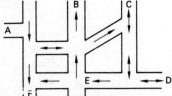

Fig. 20.28

(a) Say whether the following journeys are possible: A to C, C to A, C to B, D to A, A to E, B to any point marked.   (b) How many different routes are there from A to F?   (c) What is the maximum number of points that can be visited on one tour starting from A?   (d) What is the least change to make in the road system to make the journey D to A possible?   (e) Indicate the least arrow changes so that a motorist entering at A is trapped in the road system for ever.

Perimeters and Areas

Exercise 21

(*Use* $\pi = 22/7$ *or* 3·14.)

1   One of the runways of an airport is 1 820 m long and 32 m wide. Calculate the area of the runway in hectares.

What is the length of another runway, also 32 m wide, which has an area of 5·664 ha?

2   The lawn of a stately home measures 124 m by 46 m. It is cut twice per week and the edge trimmed once per fortnight. Calculate the area cut and the length trimmed in a month.

Find the cost of a dressing of fertilizer at 1·2p per 12 m² and a dose of weed killer at 2·4p for 20 m². Answers to the nearest penny.

3   In Fig. 21.1, the lengths are in metres. Calculate the areas P ∩ Q,   P ∪ Q   and   P ∩ Q'.

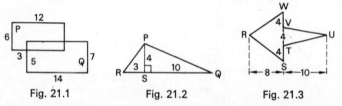

Fig. 21.1   Fig. 21.2   Fig. 21.3

4   Using Fig. 21.2, in which the lengths are cm, calculate: (a) the areas of triangles PRS, PQS and PQR,   (b) the perimeters of triangles PRS and PRQ.

5   The lengths in Fig. 21.3 are in cm. Calculate: (a) the area of the figure,   (b) the perimeter of triangle RSW,   (c) the perimeter of the whole figure.

6   A parallelogram has sides of length 12 cm and 18 cm. The distance between the 18 cm sides is 8 cm. Calculate   (a) the area of the parallelogram,   (b) the distance between the 12 cm sides.

7   A rhombus has diagonals of length 24 m and 32 m. Calculate (a) its area,   (b) its perimeter.

8   Calculate the area of a kite with diagonals of length 10 cm and 14 cm. Is it possible to find the perimeter of the kite?

9   Using 1 cm to one unit on both axes plot the trapezia listed below and calculate their areas.

(a) (0 0),   (0 7),   (1 0),   (1 11).
(b) (1 0),   (1 11),   (2 0),   (2 12),
(c) (2 0),   (2 12),   (3 0),   (3 6).
(d) (3 0),   (3 6),   (4 0),   (4 1).

Use these results to find the approximate area between the curve through (0 7),   (1 11),   (2 12),   (3 6),   and   (4 1), the x and y axes and the line $x = 4$.

10   Sketch the curve that passes through the points

x	0	1	2	3	4	5	6
y	2	$2\frac{1}{2}$	3	3	2	1	0.

Use the method of question 9 to estimate the area between the curve and the x and y axes.

11   Use the trapezoidal method of questions 9 and 10 to estimate the area between the two axes and the curve passing through the points

x	0	1	2	3	4
y	8	$7\frac{1}{2}$	6	3	0.

12   Calculate the area of the following regular polygons.
(a) a hexagon of side 4 cm,   (b) a pentagon of side 8 cm, (c) a nonagon – nine sides – of side 10 cm,   (d) an octagon drawn in a circle of radius 12 cm.

13   Plot the following polygons and calculate their areas.

(a)
x	4	4	7	6	7
y	2	6	6	4	2

(b)
x	1	2	4	6	7
y	2	4	3	4	2

(c)
x	2	2	0	-2	-2
y	-1	2	4	2	-1

(d)
x	1	4	2	-2	-4	-1
y	2	-1	-3	-3	-1	2.

14   An airliner has a turning circle of radius 2·8 km. Calculate the distance it travels in turning through 180 degrees.
A military version of the same aircraft has a radius of turn of 2·1 km. What area of countryside does it encircle in one complete turn?

15   The blades of a helicopter have a span of 4·2 m and rotate at 110 revolutions per minute. Calculate the tip speed of the blades in km/h.

16   The diameter of the wheels of a car is 56 cm. How many times do these rotate in a journey of 90 km?
How many for those of a motor scooter's wheels of diameter 21 cm?

70

17 A cyclodrome has semi-circular ends of inner radius 52·5 m. Calculate the distance around one of these ends (a) at the inner edge, (b) 3·5 m outside the inner edge, (c) 10·5 m outside the inner edge.

18 Figure 21.4, shows a running track with 90 m straights and ends of inner radius 35 m. The track is $3\frac{1}{2}$ m wide.

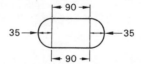

35 ──► 35

Fig. 21.4

Calculate: (a) the distance around the inner edge of the track, (b) the distance around the outer edge, (c) the area enclosed by the track, (d) the area of the track.

Geometrical Properties

Exercise 22

Angle properties

1 Express in degrees the smaller angle between the directions (a) N.E. and S 10° W, (b) N 35° W and S 18° E. (c) S.E. and N 80° W, (d) 025° and 250°, (e) 118° and 316°, (f) 032° and 277°.

2 A ship steers N 40° E for 8 nautical miles and then due East for a further 8 miles. What is the bearing of the ship from its starting point?

3 A ship sails S 30° W for 6 nautical miles and then N 30° W for a further 6 miles. How far is it then from its starting point?

4 State the smaller angle between the minute hand and the hour hand of a clock at 2, 3, 7, 8 and 11 o'clock.

5 (a) Through what angle does the hour hand of a clock turn in one hour?
   (b) State the angle between the hands of a clock at 2.30; 6.45; 7.15 and 8.20.

6 In Fig. 22.1, POQ and ROS are straight lines. Calculate the

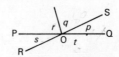

Fig. 22.1

remaining angles if (a) $\angle p = 36°$ and $\angle r = 74°$, (b) $\angle q = 51°$ and $\angle s = 48°$, (c) $\angle r = 31°$ and $\angle t = 134°$, (d) $\angle p = 47°$ and $\angle q = 55°$, (e) $\angle r = 56°$ and $\angle t = 126°$.

7 These questions refer to Fig. 22.2 in which LN is parallel to PR and MR bisects angle NMQ.
(a) If $\angle NMQ = 130°$, calculate $\angle NMR$, $\angle MRQ$, $\angle MQR$,
(b) If $\angle PQS = 64°$, calculate $\angle LMQ$, $\angle MRQ$.
(c) If $\angle QRM = 56°$, calculate $\angle PQS$.
(d) If $\angle QRM = x°$, and $\angle OMN = 3x°$, calculate x.

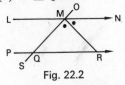

Fig. 22.2

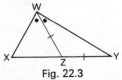

Fig. 22.3

8 In Fig. 22.3 WZ bisects $\angle XWY$ and WZ = YZ.
(a) If $\angle WYZ = 44°$, calculate $\angle XWZ$ and $\angle WXZ$. (b) If $\angle XWY = 76°$, calculate $\angle WYZ$ and $\angle WXZ$. (c) If $\angle WZX = 82°$, calculate $\angle WYZ$ and $\angle WXZ$. (d) if $\angle XWY = 90°$, show that triangle WXY is isosceles. (e) If $\angle WYZ = x°$ and $\angle WXZ = 3x°$ calculate x.

9 Figure 22.4 shows a triangle in which PQ = PR and QS = QR.

Fig. 22.4

(a) Calculate PRQ, RSQ and RQS, if QPR = 28°. (b) Calculate QSR, SQR and PQS, if QRS = 64°. (c) Calculate QRS, PQS and PSQ, if SQR = 42°. (d) Calculate x, if QPR = x° and QRS = 2x°.

10 Find the number of sides in a regular polygon with internal angles of (a) 160, (b) 165, (c) 172 and (d) 140 degrees.

11 Calculate the exterior angle and hence the interior angle of a regular polygon with (a) 10, (b) 12, (c) 15, (d) 18 and (e) 20 sides.

12 P, Q, R and S are four successive vertices in the previous polygons. Calculate (a) $\angle PRQ$, (b) $\angle PSR$ in each case. Calculate $\angle PQS$ in the final case.

13 Two of the angles of a hexagon are 90° and the other four angles are equal. Calculate the equal angles.

14   Three angles of a pentagon are equal. The other two angles are 112° and 128°. Calculate the equal angles.

15   A quadrilateral has angles in the ratio 1:2:5:4. Calculate these angles and show that the quadrilateral is cyclic.

16   Find the angles of the triangle formed by joining 1 to 6 to 8 o'clock on a twelve hour clock face.

17   Repeat question 16 for   (a) the triangle 6 to 8 to 11 o'clock, (b) the quadrilateral 1 to 3 to 7 to 9 o'clock,   (c) the quadrilateral 2 to 5 to 9 to 12 o'clock.

Properties of triangles

18   PQR and XYZ are two triangles with sides p, q, r and x, y, z. Say whether the triangles are congruent, similar or neither, in the following cases.

(a) $p = x$, $q = y$, $r = z$.   (b) $p = y$, $r = x$, $\angle P = \angle Z$.
(c) $p = x$, $\angle P = \angle X$, $\angle Q = \angle Z$.
(d) $p = x$, $\angle P = \angle Z$, $\angle Q = \angle Y$.
(e) $\angle P = \angle X$, $\angle Q = \angle Z$, $\angle R = \angle Y$.
(f) $r = z$, $q = y$, $\angle Q = \angle Y$.
(g) $r = z$, $q = y$, $\angle P = \angle X$.   (h) $r = \frac{1}{3}z$, $q = \frac{1}{3}y$, $\angle P = \angle X$.

19

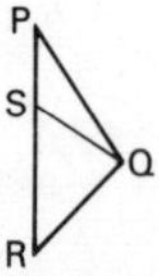

P
S
Q
Fig. 22.5   R

Make three sketches of Fig. 22.5 and show its reflexion P′ S′ R′ Q′ in   (a) PR,   (b) PQ and (c) RQ.
   In each case name three pairs of congruent triangles.

20   Refer to Fig. 22.6 with lengths in cm and calculate: (a) the length PQ,   (b) the length PR,   (c) the ratio of the areas PRS:PQS,   (d) the ratio of the areas PRS:PRQ,   (e) the ratio of the lengths of the perpendiculars from S to QP and RP.

Fig. 22.6   Fig. 22.7

21   In Fig. 22.7, the lengths are in cm. Calculate: (a) the lengths RT and ST,   (b) the area of the complete figure,   (c) the perpendicular distance from R to TS,   (d) the shortest distance from T to RQ.

22   Explain why the semi-circle drawn on the hypotenuse of a right angled triangle is equal in area to the sum of the two semi-circles on the other two sides.

23 The shadow of a vertical pole is 2·4 m long. At the same time the shadow of a fir tree is 90 m long. If the pole is 2 m high, how high is the fir tree?

24 A ladder 8·4 m long leans against a vertical wall with its base 2·1 m from the wall. How far from the wall is a point (a) 6 m from the top of the ladder,   (b) 6 m from the base of the ladder?

How far up the ladder must a bar 70 cm long be placed if it is to just touch the wall?

25 A cone 36 cm high has a diameter of 24 cm. What is the diameter of the circle formed if the cone is cut,   (a) 9 cm from the top,   (b) 12 cm from the base?

How far from the base should a cut be made to give a circle one-ninth of the area of the base?

26 Use Fig. 22.8, in which PSR = PRQ = 90°, for the following questions.   (a) Name any pair of similar triangles in the figure.   (b) Name two other pairs of similar triangles in the figure.   (c) Calculate the lengths SR, PR and PS, if the units are cm.

Fig. 22.8   Fig. 22.9

27 These questions relate to Fig. 22.9, in which PQRS is a parallelogram and the units are cm.

(a) State the lengths RQ and PQ.   (b) Explain why triangles POQ and SOT are similar.   (c) Calculate the ratios PQ:ST and PO:OT.   (d) Calculate the ratio of areas POQ:SOT, POS:POQ, POS:QSR.   (e) Deduce the ratio of the areas POS:TOQR.

Properties of quadrilaterals

28 Sketch the following, indicating any centres and any axes of symmetry: parallelogram, rectangle, rhombus, square, trapezium, isosceles trapezium, kite.

29 Draw Venn diagrams to illustrate the relation between: (a) parallelograms and rectangles,   (b) squares and rhombi, (c) parallelograms, trapezia and squares,   (d) kites (K) and parallelograms (P).

What are the elements of K ∩ P?

30 Name the most general class of quadrilaterals that has (a) both pairs of opposite sides parallel,   (b) its diagonals equal,

(c) diagonals that cut at right angles, (d) diagonals that bisect at right angles, (e) all four angles equal, (f) all four sides equal.

31 PQRS is a parallelogram and X and Y are points on QS such that $QX = YS = \frac{1}{4} QS$. Show that (a) triangles PSY and RQX are congruent, (b) triangles PQX and RSY are congruent, (c) figure PXRY is a parallelogram.

32 WXYZ is a rhombus. S, P, Q and R are the mid-points of its sides. Explain why (a) PQRS is not a rhombus, (b) why PQRS is a parallelogram.

Ratio properties
33 In Fig. 22.10, ST is parallel to QR. Calculate
  (a) PS and QR, if SQ = 6 cm, PT = 9 cm, PR = 18 cm and ST = 8 cm.
  (b) PR and ST, if PT = 18 cm, PS = 15 cm, SQ = 12 cm and QR = 24 cm.
  (c) SQ and ST if PS = 8 cm, PT = 10 cm, PR = 15 cm and QR = 18 cm.
  (d) the ratios PS:PQ, PT:TR and area PST:area PQR, if ST = 8 cm and QR = 20 cm.
  (e) the ratios area PST:area PQR and area STRQ: area PQR if ST = 9 cm and QR = 15 cm.

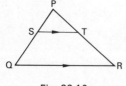

Fig. 22.10

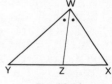

Fig. 22.11

34 For the following questions use Fig. 22.11, in which WZ is the bisector of angle XWY.
  (a) Calculate YZ if WY = 20 cm, WX = 12 cm and XY = 16 cm.
  (b) Deduce the length XZ if WY = 18 cm, WX = 15 cm and XY = 12 cm.
  (c) Calculate WX if WY = 24 cm, YZ = 18 cm and YX = 30 cm.
  (d) Find p if WY = p, WX = $p - 6$, YZ = 10 cm and XZ = 6 cm.
  (e) Find q if WX = q, WY = $q + 6$, XZ = $q - 5$ and YZ = $q - 1$.

Constructions

Exercise 23

1 Without using a protractor, construct angles of (a) 60°, 30°, 120°, 150°, (b) 90°, 45°, 135°, 67½°.

2 Draw five lines 8·7 cm long and divide these (a) into 3 equal parts, (b) into 5 equal parts, (c) in the ratio 3:1, (d) in the ratio 2:5, (e) in the ratio 31:53.

3 Construct a triangle of sides 10, 7 and 9 cm. Bisect the sides at right angles. Hence draw the circumcircle of the triangle. Measure its radius.

4 Three power boats are located 9, 7 and 6 km apart. They are to race to a point P equidistant from each of them. Construct a triangle to represent the relative positions of the boats and locate P. How far do the boats race? Explain how to position a fourth boat so that it is just as far from P as the others.

5 Construct a triangle of sides 7, 9 and 11 cm. Bisect its angles internally. From the point where the bisectors meet construct a perpendicular to the 11 cm side. Hence construct the inscribed circle of the triangle.

6 Draw a circle of radius 2·5 cm. Construct three radii of the circle with angles between them of 90°, 135° and 135°.
  Construct the tangents at the ends of these radii and so form a triangle. Measure its angles. Comment on the angles and on the symmetry of the figure.

7 In a circle of radius 6·6 cm, construct a triangle with its vertices on the circle and two of its sides 9 cm and 7 cm. Measure the third side.
  Construct the reflexion of the figure in the third side.

8 Draw four circles of radius 5 cm. Construct the following regular polygons with vertices on the circle.
  (a) a hexagon, (b) an octagon, (c) a decagon – 10 sides, (d) a 15-sided polygon.
  Mark and count the axes of symmetry on each polygon.

9 Draw a circle of radius 6 cm. Construct a regular pentagon in this circle. Now construct a pentagonal star by joining each vertex to the next vertex plus one, i.e. by the mapping, $v \rightarrow v + 2$.

10 From a point P construct three lines PQ, PR and PS each

of length 2 cm with angles between them of 135°, 120° and 105°. Join PQR to form a triangle.

Using P as the centre of enlargement, construct two further triangles similar to triangle PQR with sides two and three times as long as the original. What is the ratio of the areas of these triangles?

11 PQRS is a quadrilateral with PQ = 10 cm, angle PQR = 90°, QR = 4 cm, RS = 5 cm and PS = 7 cm. Construct the quadrilateral and locate the point which is equidistant from P and Q and just as far from side RS as from side RQ.

12 Construct a semi-circle of radius 5·6 cm. Shade the set of points within the semi-circle that are nearer to the circumference than to the centre.

13 Draw any convex polygon – i.e. one with all its angles less than 180°. Bisect its sides and join these four mid-points. Comment on the resulting figure.

14 Repeat question 13, using a concave quadrilateral – i.e. one with an angle greater than 180°.

15 *Use graph paper and a scale of 1 cm to 1 nautical mile to make scale drawings for the following.*

(a) A yacht sails on a course of 052° for 4 n miles and then 145° for 7 n miles. Use your drawing to find the distance of the yacht from its starting point and the course to steer to return to that point.

(b) A ship steams at 15 knots on a course of 033° for 20 minutes. It then steams at the same speed on a course 170°. From your scale drawing find (a) the distance the ship travelled before it was due East of its starting point, (b) after what time it was South West of its starting point.

(c) Lighthouse L_1 is North East of lighthouse L_2 and 8 n miles away. Ship S is 4 n miles from L_1 on a bearing 130°. From your drawing find (a) the distance of S from L_2, (b) the bearing of S from L_2.

16 Three trees T_1, T_2 and T_3 are growing with distances $T_1T_2 = 75$ m, $T_2T_3 = 85$ m and $T_3T_1 = 100$ m. A fourth tree is to be planted equidistant from each of the other three. Find, using a suitable scale drawing, the position of T_4 and the distance T_1T_4.

17 Mark two points P and Q, 6 cm apart. Find the pair of points 3 cm from P and 6 cm from Q and the pair 4 cm from P and 8 cm from Q. Add further pairs of points twice as far from Q as from P and sketch the locus of these points.

18 Construct the following loci.
(a) Points equidistant from a given point. (b) Points equidistant from two given points. (c) Point(s) equidistant from three given points.

Is it possible to have a point equidistant from four given points?

19 A plank of wood 6 m long is placed vertically against a wall and gradually slides down with one end touching the wall until it is horizontal. Using graph paper and compasses with a scale of 2 cm to 1 m, construct positions of the plank at $\frac{1}{2}$ m intervals. Hence plot the locus of the mid-point of the plank.

20 Repeat question 19, (a) finding the locus of the point 2 m from the top of the plank, (b) finding the locus of the point $1\frac{1}{2}$ m from the base of the plank.

Areas and Volumes

Exercise 24

$$(Use\ \pi = \frac{22}{7}\ or\ 3\cdot14)$$

Prisms and cylinders

1 Calculate the volumes of the prisms shown in Fig. 24.1. All lengths are in cm.

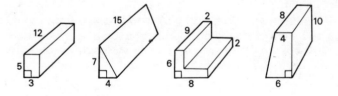

Fig. 24.1

2 Calculate the volumes of the 'letter' prisms shown in Fig. 24.2. The lengths are in cm and each part of the letters is 2 cm wide and 2 cm thick.

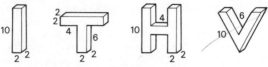

Fig. 24.2

3 An open cardboard box measures 30 by 20 by 14 cm. Calculate (a) its volume, (b) the area of cardboard in the box.

4 Sketch nets of (a) a cube of side 1 cm, (b) a cuboid with sides 3, 2 and 1 cm, (c) a triangular prism with the sides of the triangle $1\frac{1}{2}$, 2 and $2\frac{1}{2}$ cm and length 10 cm.
Calculate the surface area of each prism.

5 The cuboid shown in Fig. 24.3a has lengths in cm. It is cut into two halves through BC and EH.
(a) Calculate the area of one face of the cut.
(b) Explain how the two halves can be fitted together to form Fig. 24.3b.
(c) Sketch the other two solids that can be formed by (1) joining F to D and G to A, (2) joining A to G and E to C.

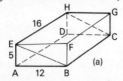 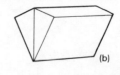

Fig. 24.3

(d) Test the formula $F + V = E + 2$, for each of the four solids.
(e) Which one of the solids has the same surface area as the original?
(f) Calculate the increase in surface area in the other two cases.

6 Calculate the volume of (a) wood in a rolling pin 42 cm long and 4 cm in diameter, (b) lemonade in a lemonade-filled straw 4 mm in diameter and 21 cm in length, (c) toffee in 120 000 toffee rolls diameter 1·8 cm and length 1·4 cm.

7 'SPICK' washing up fluid comes in containers of radius 4 cm and length 8 cm. 'SPAN' fluid is packed in cylinders of radius 3 cm and length 14 cm. Which container holds more fluid?

8 Calculate the volume of a cylinder radius 7 cm and height 10 cm.

With minimum working, calculate the volume of the cylinder with (a) half the height, (b) half the radius, (c) half the height and half the radius.

9 Find the volume of (a) water in 2 km of water main with diameter 1 m, (b) oil in 400 km of pipeline with diameter 42 cm, (c) mercury in 12 cm of tube, diameter 1 mm.

10 WXYZ is a rectangle with WX = 4 cm and WZ = 8 cm. Calculate the volume of the cylinder formed when the rectangle is rotated through 360° about (a) side WX, (b) side WZ, (c) the line joining the mid points of WX and ZY.

11 A tunnel, 240 m long, is constructed in the form of a half cylinder of radius 4·2 m. Calculate the volume of earth removed in its construction.

12 Repeat question 11 for a tube tunnel $1\frac{1}{2}$ km long with cross section a 270° sector of a circle of radius 5·6 m.

13 Figure 24.4 shows a cylinder, diameter 4 cm, that has been sliced off at one end at an angle of 45°.
(a) Sketch how two such sliced cylinders could be fitted together to form one complete cylinder.
(b) Use your sketch to help you to find the volume of the sliced cylinder.

Fig. 24.4

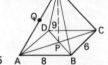

Fig. 24.5

Pyramids and cones

14 OABCD, Fig. 24.5, is a pyramid with lengths as shown in cm. P is the mid point of diagonal AC and Q is the mid point of edge OA.
Calculate the volumes of the pyramids, OABCD, OADC, OABP, OADP and QABD.

15 ABCDEFGH is a cuboid with AB = 15 cm, BC = 8 cm and CG = 6 cm.
(a) Calculate the volumes of the pyramids ABDH and BCGA, ABCDG and BCGFH. Comment on the relation between the answers.
(b) If O is the point where the diagonals AG and EC cut, find the volume of pyramid OABFE. How many other pyramids in the figure have the same volume?

16 The triangle LMN has sides LM = 12 cm, MN = 9 cm and angle LMN = 90°. Calculate the volumes of the cones

80

formed when the triangle is rotated through 360° about   (a) LM   (b) MN,   (c) LN.

17   A double cone is fitted in a cylinder of radius 4 cm, height 14 cm – see Fig. 24.6. Calculate the volume of the double cone.

Fig. 24.6   Fig. 24.7

18   The flower trough shown in Fig. 24.7 is part of a cone. It has an upper diameter of 90 cm and a lower diameter of 40 cm. It is 30 cm deep. Calculate   (a) the height of the cone of which the trough is part,   (b) the volume of soil needed to fill the trough.

Spheres

19   Calculate the volume of   (a) ivory in a billiard ball radius 2·1 cm,   (b) iron in a cannon ball of radius 14 cm,   (c) air in a football of radius 10·5 cm,   (d) medicine in a medicine ball of diameter 35 cm!

20   Take the radius of the earth as 6 400 km and calculate (a) its volume,   (b) the land area – assuming five sixths of the surface is sea.

21   A cylinder and a sphere both have equal volumes and both have a radius of 12 cm. What is the height of the cylinder?

22   Find the difference in volume between a cube of edge 6 cm and a sphere of diameter 6 cm.

23   Which has the greater surface area, a sphere of radius 4 cm or a hemi-sphere of radius 5 cm?

24   Marbles of diameter 1 cm are dropped into a beaker of water, diameter 4 cm. How many marbles are dropped in if the water rises 2 cm?

25   On close examination, a skinless sausage proves to be a cylinder of radius 1·4 cm, length 8 cm with two hemispherical ends. Calculate the volume of sausage meat in a factory's output of 400 000 bangers.

Angles and Diagonals

Exercise 25

Use Fig. 25.1, in answering the following questions. The lengths are in cm.

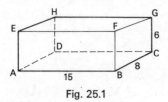

Fig. 25.1

1   Name three angles equal to   (a) ACE,   (b) AHB, (c) EBG.

2   Name three diagonals equal in length to   (a) AH,   (b) HC, (c) CE.

3   Name three triangles congruent to triangles ABC, ADH, EAC and ABG.

4   Classify the following angles as acute, obtuse or right angled. ∠EAD, ∠EDA, ∠EGC, ∠BHD, ∠AGB, ∠DGB.

5   Arrange the following diagonals in order of increasing length. AG, AH, HC and EG.

6   Which point is furthest from corner A?

7   Is the angle between diagonals HB and FD larger, smaller or equal to the angle between AG and EC?

8   Is it possible to construct a sphere to pass through the eight vertices AB . . . H? If so where is its centre?

9   Arrange the following rectangles in order of increasing area. ABCD, ABFE, BCGF, ABGH, HFBD.

10   Calculate angles FAB, HAD and GAC.

11   Calculate the lengths AC, AH and AG.

12   Calculate the angles AHB and GAF.

13   What is the shortest distance from A to G,   (a) along the edges of the cuboid,   (b) across its surface,   (c) by any route.

14   ABCDEFGH is a cube. Say whether the following triangles are isosceles or scalene, right angled or not right angled. Triangles EFH, HAB, BDE and AGF.

15 Repeat question 14 for (a) a cuboid with ABCD square, (b) a cuboid with no faces square.

16 A flagpole, 24 m high stands at one corner of a rectangular parade ground with sides 50 m by 120 m. Calculate the angle of elevation of the top of the flagpole from the other three corners of the ground.

17 A television mast is 240 m high. Calculate the angle of elevation of its top from (a) a point P 200 m due East, (b) a point Q 150 m due South, (c) the point R on PQ nearest to the foot of the mast.

These questions refer to a cuboid ABCDEFGH.

18 If AB = 12 cm, AE = 5 cm and AD = 9 cm, calculate: (a) lengths AF, AC and AG, (b) angles FAB, HAD and GAC, (c) the angle between diagonals AG and BH.

19 If AB = 12 cm, AE = 3 cm and AD = 4 cm, calculate: (a) lengths AH, AC and AG, (b) angles HAD and GAC, (c) the angle between CE and FD.

20 If AB = p, BC = q and CG = r, express in terms of p, q or r the lengths of the diagonals AC, CF, FA and AG.

21 Use the final result of question 20 to find the length of the longest diagonal in cuboids with sides, (a) 1, 2, 2 cm, (b) 2, 3, 6 cm, (c) 4, 4, 7 cm, (d) 2, 6, 9 cm, (e) 4, 8, 8 cm.

22 The eight vertices of a cuboid lie on a sphere. Calculate the radius of the sphere if the sides of the cuboid are (a) 8, 4 and 1 cm, (b) 6, 6 and 3 cm, (c) 7, 6 and 6 cm.

These questions refer to the pyramid VABCD Fig. 25.2, with height VO = 12 cm, sides AB = 8 cm and BC = 6 cm. M is the mid-point of AB.

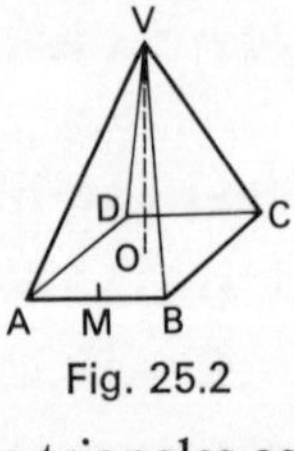

Fig. 25.2

23 Classify the following triangles as isosceles or scalene, right angled or not right angled. Triangles VAB, VAC, VMC and VMB.

24 Name a triangle congruent to triangle VAB, VBC, ADC and VDB.

25 Calculate the lengths AC, AV and MV.

26 Calculate the angles CAB and VAC.

27 Calculate the angles between (a) the face VAB and the base, (b) the edge VA and the base.

28 Repeat questions 23 to 27 for AB = BC = 8 cm and OV = 7 cm.

The next two questions concern a sphere centre O *and radius* 10 *cm.*

29 Calculate the area of circles (a) 6 cm, (b) 7 cm and (c) 9 cm from O. Leave π in your answers.

30 Calculate the area of circles that subtend angles of (a) 120°, (b) 90°, (c) 60° at O. Leave π in your answers.

31 Two intersecting spheres have their centres 10 cm apart. The radius of one sphere is 6 cm and that of the other 8 cm. Calculate the radius of the circle in which they intersect.

32 Repeat question 31 for (a) two spheres radii 5 and 12 cm with centres 13 cm apart, (b) two spheres radii 9 and 20 cm with centres 29 cm apart.

Answers

Exercise 1

1 (a) {M, T, W, Th, F, S, Su}
   (b) {Jan, March, May, July, Aug, Oct, Dec.}
   (c) {}   (d) {19, 23}   (e) {6, 12, 18 . . .}
   (f) {2, 3, 5, 6, 10, 15, 30}; e, c.

2 (a) {vowels}   (b) {colours of traffic lights}
   (c) {last 5 letters of the alphabet}
   (d) {multiples of 23}   (e) {polygons}; d and e.

3 (a) finite   (b) infinite   (c) finite   (d) null
   (e) finite   (f) infinite   (g) infinite.

4 {A, B, C, D, E, F, G};   {B, D};   {A, F, G};
   {C, E}. $R \cup S = (R \cap S) \cup (R \cap S') \cup (R' \cap S)$.

5 {Th, F, S, Su};   {M, T, W, Th};   {T, W};
   {M, Th, F, S, Su};   {M, Th, F, S, Su}.
   $M' \cup T' = (M \cap T)'$.

6 (a) 3, 6, 9;   5, 10, 15;   1, 4, 9;   1, 2, 3
   (b) 15, 30, 45   (c) 3, 5, 6   (d) 1, 3, 4
   (e) 9, 36, 81   (f) 225, 900, 2 025.

9 (a) T   (b) T   (c) F   (d) F   (e) T.

10 (b) $\subset$ (c) $\supset$ (f) $\supset$.

11 (a) 3   (b) 3   (c) 7.

12 (a) 4   (b) 6   (c) 4   (d) 14.

13 (a) {x, y}   {x}   {y} ϕ
   (b) {r, s, t}   {r, s}   {r, t}   {s, t}   {r}
   {s}   {t} ϕ.   4, 8.

14 16, 32.   **15** 64, 128, 256, 1 024. $2^0 = 1$.

16 (a) 8 subsets   (b) 9 subsets   (c) 4 subsets
   (a) 9   (b) 9   (c) 16.

17 23, 17, 7, 33, 15, 28.

18 14, 16, 5, 25, 17, 20.   6, 14, 6, 14, 12, 10.

19 27, 17, 7, 43, 11, 16, 26.   **20** 16, 4, 4, 1.

21 7, 7, 1, 13, 28, 21.   **22** 5, 30, 15, 10, 30.

25 (c) Note goalkeepers in hockey also
(e) Note 2 is an even prime.

29 (a) T (b) T (c) T (d) F (e) T (f) T
(g) F.

30 (a) $E \cap O = \phi$; $S \subset T$ (b) 6, 12, 18, 24
(c) $E \cap T = S$ (e) $E \cap O$; $S \cap O$.

31 (b) Bicycles not made for two, non veteran bicycles, bicycles that are neither veteran nor made for two.

32 8. **33** 46. **34** 11, 1.

35 14. **36** 10.

37 (a) $2n$ (b) $(26 - 2n)$, $(31 - 2n)$ (c) 9 (d) 13.

38 {April, May, June, July, Aug, Sept, Oct} – both hotels open
{March to Dec inclusive} – at least one hotel open
{Jan, Feb, March, Dec} – hotel T closed
{March, Dec} – T closed but S open.

39 Days when at least one baker calls
Days when both call
Days when B does not call
Days when neither C nor B calls
Days when B but not C calls.

40 98, 32.

Exercise 2

1 (a) 88 502 (b) 3 053 (c) 14 000 006 (d) 404 605.

2 (a) 495 (b) 5 940 (c) 6 993
(d) 79 992, 79 200, 792 (e) 891, 810, 81.

3 {361, 610, 519} The first average is half the second.

4 {123, 122, 245}.

5 {26, 30, 36, 32, 38, 42}; {71, 83, 116, 104, 137, 149};
{203, 294, 449, 361, 516, 607}.

6 The first average is half the second.
Each number is counted twice in the second case.

7 {36, 42, 48, 54, 60, 66, 72, 78, 84, 90, 96};
{32, 40, 48, 56, 64, 72, 80, 88, 96};
{common multiples of 6 and 8}.

8 (a) (5, 19)   (7, 17)   (11, 13)
(b) (3, 37)   (11, 29)   (17, 23)   (c) (2, 41)
(d) (2, 53)   (e) (7, 73)   (13, 67)   (19, 61)   (37, 43).

9 (a) (11, 3)   (13, 5)   (19, 11)
(b) (17, 5)   (19, 7)   (23, 11)
(c) (29, 3)   (31, 5)   (37, 11)
(d) (37, 3)   (41, 7)   (47, 13).
No solutions since either p or q is even and hence not prime.

10 (a) (0, 20)   (1, 19)   (2, 18)   ...   (10, 10)
(b) (1, 24)   (2, 12)   (3, 8)   (4, 6)
(c) (1, 1, 144)   (2, 2, 36)   (3, 3, 16)   (4, 4, 9)
(6, 6, 4)   (12, 12, 1)
An infinite number of pairs.

11 (a) 162, 168, 174, 180, 186, 192, 198
(b) 161, 168, 175, 182, 189, 196
(c) 160, 168, 176, 184, 192
168;   168 and 192;   168.

12 1   4   9   16   25   36   49   64   81   100
1   8   27   64   Yes, 64.

13 (a) 567   765   (b) 248   842
(c) 335   533   198   594   198.

14 49   102   155.

15 (a) O   (b) E   (c) E   (d) O   (e) O   (f) E
(g) E   (h) O.

16 121   144   169   196   225
(a) 855   (b) 171   (c) 23   25   27   29
256   289   324   361   400.

17

3	1	1	1	1	1	1
4	1	0	1	0	1	0
5	1	1	1	1	1	1
7	0	0	1	1	0	1

15	1	1	1	1	1	1
28	0	0	1	0	0	0
35	0	0	1	1	0	1
60	1	0	1	0	1	0

18 (a) 4   168   (b) 18   108   (c) 12   336
(d) 26   364   (e) 15   900   (f) 22   1 452.

19 (a) 11   (b) 101   (c) 1 009.

20 (a) 5   (b) 525   (c) 8.   **21** 20·18 h.

22 (a) 20·12 h   (b) 20·21 h   (c) 20·14 h.

23 (a) 15   (b) 201.

24 (a) 79   (b) −11.

25 (a) 60   375   (b) 20   125   (c) 23   625.

26 (a) 27 too large   (b) 18 too large
(c) 648 too large   (d) 1 305 too large.

27 (a) 1   3   6   10   15   21   (b) 28   36   78.

28 (a) 382+   (b) 497−   (c) 423×   (d)

$$
\begin{array}{r}
3\cdot5 \\
83\overline{)290\cdot0} \\
249 \\
\hline
41\ 5 \\
41\ 5 \\
\hline
00\cdot0
\end{array}
$$

(a)
$$
\begin{array}{r}
382 \\
265 \\
\hline
647
\end{array}
$$

(b)
$$
\begin{array}{r}
497 \\
318 \\
\hline
179
\end{array}
$$

(c)
$$
\begin{array}{r}
423 \\
81 \\
\hline
33840 \\
423 \\
\hline
34263
\end{array}
$$

29 (a) 17   21   (b) 16   36   (c) 26   37
(d) 162   1 458   (e) 55   91   (f) 26   42.

Exercise 3

2 $\frac{5}{6}$ $\frac{2}{3}$ $\frac{7}{8}$ $\frac{5}{8}$ $\frac{7}{12}$ $1\frac{1}{2}$.

6 (a) $\frac{10}{15} = \frac{32}{48} = \frac{50}{75} = \frac{2000}{3000}$ (b) $\frac{20}{32} = \frac{35}{56} = \frac{45}{72} = \frac{250}{400}$

(c) $\frac{18}{24} = \frac{21}{28} = \frac{27}{36} = \frac{120}{160}$ (d) $\frac{35}{45} = \frac{63}{81} = \frac{70}{90} = \frac{777}{999}$

7 (a) $\frac{3}{5}$ (b) $\frac{2}{3}$ (c) $1\frac{1}{12}$ (d) $\frac{5}{9}$ (e) $\frac{3}{8}$ (f) $\frac{7}{12}$.

9 $\frac{7}{12}$ $1\frac{17}{24}$ $6\frac{11}{20}$ $\frac{2}{15}$ $\frac{1}{36}$ $2\frac{7}{15}$ $1\frac{8}{15}$ $4\frac{5}{12}$ $2\frac{11}{24}$.

10 $\frac{1}{9}$ $\frac{1}{8}$ $\frac{1}{2}$ $\frac{3}{4}$ $\frac{1}{2}$ $1\frac{3}{4}$ $14\frac{2}{3}$ 10.

11 5   48   10   2 $\frac{2}{3}$ $6\frac{2}{3}$ $1\frac{1}{5}$ $1\frac{1}{5}$.

12 3   3   3   5 $1\frac{2}{3}$ 1.

13 $3\frac{5}{6}$ $3\frac{1}{2}$ $1\frac{5}{9}$ $\frac{5}{21}$.

14 $\frac{13}{20}$ $8\frac{9}{20}$ $1\frac{1}{4}$ 9 $1\frac{9}{16}$.

15 50   25 $12\frac{1}{2}$ 60   70   45   64   62   98 p.

16 12   50   45   18   35   40 $37\frac{1}{2}$ mins.

17 $\frac{1}{14}$ $\frac{5}{24}$ $\frac{1}{3}$ $\frac{5}{21}$ $\frac{3}{8}$.

18 $\frac{1}{15}$.   **19** £21.   **20** $\frac{1}{16}$.

21 $\frac{1}{2}$ $\frac{1}{8}$ $\frac{1}{2}$ $\frac{1}{4}$ $\frac{1}{8}$.   **22** $10\frac{1}{2}$ min.

23 60 min.   **24** 90 s.

Exercise 4

1 $1 + \frac{3}{10} + \frac{2}{100}$; $2 + \frac{4}{10} + \frac{6}{100}$; $3 + \frac{5}{100} + \frac{2}{1000}$;
 $6 + \frac{7}{10} + \frac{8}{1000}$; $9 + \frac{4}{10} + \frac{5}{10000}$;
 $5 + .\frac{3}{100} + \frac{2}{10000}$.

2 8·3   9·07   7·26   3·075   8·203.

3 2·7   3·05   4·29   12·81   9·192   14·031   10·206.

4 36·2   2 980   628   0·45   0·153   0·00158   0·0837.

5 0·09   0·64   0·25   0·0004   0·000064   0·2   0·3
0·4   0·05.

6 21·40   25·23   36·96   22·33   13·02.

7 (a) 14·12   13·23   20·96   10·89   6·02
(b) 2   0·1   0·3   1·5   2·1.

8 20·3   11·76   47·04   1·96   6   24   4   7·84
0·343   282·24.

9 36·9   12·18   37·02   36·78   5·904   605·16   410
4   102·5   0·009756   0·0144   151·29.

10 (a) 34·46   18·32   5·06   0·04
(b) 34·5   18·3   5·1   0·0
(c) 34·5   18·3   5·06   0·0382
(d) 30   20   5   0·04.

11 0·3̇3   0·6̇6̇   0·11̇   0·22̇   0·44̇   0·45̇   0·6̇3̇
0·8̇1̇   0·5̇71428̇   0·7̇14285̇   0·8̇57142̇.

12 (a) 0·333   0·667   0·111   0·222   0·444   0·455
0·636   0·818   0·571   0·714   0·857
(b) 0·33   0·67   0·11   0·22   0·44   0·45   0·64
0·82   0·57   0·71   0·86.

13 $\frac{4}{33}$ $\frac{31}{99}$ $\frac{6}{11}$ $\frac{2}{11}$ $\frac{2}{3}$ $\frac{48}{111}$.

14 $\frac{3}{4}$ 0·75; $\frac{11}{20}$ 0·55; $\frac{18}{25}$ 0·72; $\frac{7}{8}$ 0·875;
$\frac{12}{25}$ 0·48; $\frac{17}{20}$ 0·85; $\frac{1}{40}$ 0·025; $1\frac{1}{5}$ 1·20;
$2\frac{13}{20}$ 2·65.

15 50% 0·5;   25% 0·25;   75% 0·75;
$12\frac{1}{2}$% 0·125; $37\frac{1}{2}$% 0·375; $62\frac{1}{2}$% 0·625;
$87\frac{1}{2}$% 0·875;   10% 0·1;   30% 0·3;   70% 0·7;
15% 0·15;   45% 0·45;   65% 0·65.

16 $\frac{7}{10}$ 70%; $\frac{16}{25}$ 64%; $\frac{7}{20}$ 35%; $\frac{3}{8}$ $37\frac{1}{2}$%;
$\frac{19}{20}$ 95%; $1\frac{9}{20}$ 145%; $2\frac{18}{25}$ 272%; $3\frac{3}{20}$ 315%.

17 £16·80   £11·20   £8·40   £7·70   £6·65   £4·58$\frac{1}{2}$
67p   38p   70p.

18 94p   78p   62p   42p   (a) 75p   (b) £2·00.

19 (a) £7 (b) £5·85 (c) £7·02 (d) £14·84 New.
20 20% $13\frac{1}{3}$% 10% 8% $5\frac{5}{9}$% $4\frac{1}{6}$%.

Exercise 5

1 £3·30 £3·96 £5·28 £6·16 £7·92.
2 The final two.
3 £7·81 £8·76 £9·52 £10·66 £15·41.
4 No. **5** Gold.
6 £23·75 £10·40 £17·75 £33·50.
7 £1·89 £3·35$\frac{1}{2}$ £5·42 £9·82 £10·42.
8 £7·20 £7·90 £9·42 £9·44 £13·96 £12·87.
9 £36 £151·20 £6·37$\frac{1}{2}$ £4·20 £197·77$\frac{1}{2}$.
10 £19·78 £29·67 £24·72.
11 £48 £60 £64 £84·80 £96.
12 (a) £51 £63·75 £68 £90·10 £102
(b) £52·50 £65·60 £70 £92·75 £105.
13 72p 80p; 32.
14 (a) 84p (b) £1·80 (c) £3·60.
15 (a) Nil (b) 36p (c) 72p.
16 (a) £1·80 (b) 52p (c) £6·16 (d) £2·40.
17 (a) £1·05 (b) £2·31 (c) £2·35 (d) £5·71.
18 (a) 7 (b) 19.
19 (a) £105 (b) £108·80 (c) £112·50 (d) £287·50
(e) £56·70.
20 (a) $2\frac{1}{2}$% (b) $4\frac{1}{6}$%.
21 (a) 5 years (b) 8 years.
22 (a) £300 (b) £300.
23 (a) £49·95 (b) £38·20 (c) £81·41 (d) £31·07.
24 £4 750. **25** £5 800 £31 850. **26** 410.
27 £440 £650 £820 £1 010.
28 £220 £280 £410 £690.
29 £431·65. **30** £42·38 £1 426·95.
31 12% $12\frac{1}{2}$% £42 £55·20 £45 £60.
32 $12\frac{1}{2}$ £36·80.
33 (a) £2·25 (b) 45p (c) 20%. **34** 25%.

Exercise 6

1 (a) 24p   (b) 39p   (c) 72p   (d) £1·14   (e) £7·50.

2 £9   £14·75   £20   £28   £36.

3 £150   £135   £195   £315   £412·50.

4 £42·50   £6·25   £15   £600.

5 £40   £70   £55   £105   £125   £140   £157.

6 £38·40   £67·20   £52·80   £100·80   £120
£134·40   £150·72.

7 54p   60p   75p   84p   105p   127½p.

8 45p   50p   62½p   70p   87½p   106¼p.

9 £11·76   £17·82   £14·64   £14·10.

10 24p   36p   29p   28p.

11 £81   £84·20   £95   £100   £111·25.

12 £750   £2 000   £2 400.

13 £97·60   £98·80   £99·85   £103·60.

14 £40   £85·76   £106·88   £117·12.

15 £2 700   £2 400.   **16** £2·43.

17 £2·59   £2·82.   **18** £60   £67   £81·40.

19 £2·31   £2.40   £2·46   £2·62.   **20** £3·52.

21 (a) A   (b) B   (c) B   (d) B.

22 (a) £8·50   (b) £98·50   (c) £13·50.

23 £3   £3·40.

24 (a) £33·25   (b) £7   (c) £36·70   (d) £3·45.

25 £13·88   £72·80   £50·10   £128·40.

26 (a) £3 680   (b) £23   (c) £1 067   (d) £3·45.

27 (a) £1 813   (b) £960   (c) £416.

28 (a) 75p   (b) £5·64   (c) £35·72.

29 (a) £2·240   (b) £188 160   (c) 2·5p   (d) £2·70.

30 £4·20   £4·56   £5·10   £6·12   £6·96.

31 £120.   **32** £192   £264   £318   £525.

33 (a) £1175   (b) £785   (c) £235·50.   **34** (a) £97·50
(b) £388·50   (c) £402   (d) £241·50   (e) £226·50.

Exercise 7

1 (a) 800   800   9 100   500   1 000   700   1 900   0
   (b) 800   850   9 100   500   1 050   700   1 900
   50.

2 (a) 13·64   5·75   0·09   359·95   67·28   183·38
   555·56
   (b) 13·6   5·75   0·0865   360   67·3   183   556.

3 (a) 506·836   506·84   506·8
   (b) 507   510   500.

4 (a) 1·73   0·17   0·02   0·00
   (b) 1·7   0·17   0·017   0·0017.

5 3·14   3·142   3·14   3·1   3.

6 $36 \pm 1·5$; $17 \pm 0·5$; $5 \pm 0·5$; 18 ± 1;
$144·25 \pm 12$; $196·25 \pm 14$.

7 44 ± 1; 4 ± 1; $112 \pm 2·5$; $32 \pm 2·5$;
458·25 to 502·24.

8 20   20   360   8 000   700.

9 314   940   110   220   200   750   70.

10 1 600   35 000   200   0·7   5   17.

12 (a) 234   2·34   23 400   0·0234
   (b) 8·07   1 470   234 000   0·341
   (c) 123   49 300   11·1   0·197   0·00308.

13 (a) 7·60   2·40   24·0   0·760
   (b) 4·01   5·12   19·1   0·681   0·238.

14 22·0   102   7·22   0·421.

15 (a) 0·435   0·0435   4·35   435
   (b) 0·0641   0·641   6·41   0·00641.
   (c) 2·50   0·136   0·108
   (d) 5·50.

16 $\dfrac{1}{77}$ $\dfrac{1}{7·7}$ $\dfrac{1}{7}$ $\dfrac{1}{0·77}$.

17 1·27   2·446   0·754 $\bar{1}·215$ 3·458 $\bar{2}·863$ $\bar{3}·928$.

18 2·70   20·6   320   0·032   1 830   0·00183   1·83.

19 (a) 2·07   207   0·0207   20 700
   (b) 463   0·463   0·00463   4 630
   (c) 33·5   0·0335   335   0·00000335.

20 (a) 0·4   1·2 $\bar{1}·2$ $\bar{1}·8$
   (b) 0·1   0·7 $\bar{1}·1$ $\bar{1}·9$
   (c) $\bar{3}·8$ 1·2 $\bar{2}·5$ $\bar{1}·75$
   (d) $\bar{3}·3$ 1·7 $\bar{2}·0$ $\bar{2}·9$.

21 (a) 110 000   (b) 2·06   (c) 0·485   (d) 1 250
    (e) 3·64   (f) 27·2   (g) 0·475   (h) 1·46
    (i) 14·6   (j) 3·41   (k) 7·35   (l) 0·735
    (m) 11·6   (n) 2·63   (o) 0·653   (p) 3 800
    (q) 3 800 000   (r) 0·00380   (s) 20·6   (t) 0·0764
    (u) 0·380   (v) 0·122ₛ   (w) 12·2   (x) 765
    (y) 2·60   (z) 11 600.

22 1 150   2·06   2 370   4·86   0·0629.

23 138   4·76   28·9   5·06   0·411.

24 $\dfrac{m}{l}$ lm $\dfrac{l}{m}$ $\dfrac{l}{m}$ l^2.

25 109   127   150   163   170   178.

26 (a) £1·94   £2·26   £2·90   £3·66
    (b) £2·90   £3·39   £4·35   £5·48.

27 £1·11   £1·66½   £1·85   £2·03½   £2·22.

28 8   16½   24½   32½   40½   49   57   65   73½
    81½p.

29 53·3   70·3   178   38·1   3·81   302 cm   227
    394   2 530   115   1·15   7 240 cm².

30 14·1   15·6   19·3   21·0   32·6   38·3.

Exercise 8

1 (a) 6:1   (b) 1:4   (c) 7:18   (d) 2:9
    (e) 5:11   (f) 1:8   (g) 5:61   (h) 6:73.

2 (a) 30   20   10 min   (b) 25   15   20 min
    (c) 24   8   28 min   (d) 27   21   12 min.

3 18   30   42 cm   No.

4 10:7:11;   8:6:9.

5 (a) £45   £105   (b) £80   £70   (c) £48   £102
    (d) £82·50   £67·50   (e) £60   £50   £40
    (f) £60   £37·50   £52·50   (g) £78   £48   £24.

6 (a) 2:25   (b) 1:20.

7 (a) 9:14   (b) 2:3   (c) 3, 7 kg   (d) $\frac{1}{15}$ $\frac{1}{10}$.

8 125   62·5   312·5 kg.

9 (a) 600   400   1 400 kg   (b) 600   200   800 kg
    (c) 800   600   2 000 kg.

10 (a) 20°   60°   100°   (b) 45°   60°   75°
     (c) 54°   60°   66°   (d) 100°   50°   30°
     (e) 40°   65°   75°.

11 (a) £1 600   (b) £240   (c) £1 440.

12 £5 400.

13 (a) 1 500   (b) 2 000 pieces of eight.

14 $13\frac{1}{2}$ $7\frac{1}{2}$ years.   **15** 99 min.

16 45 mins $67\frac{1}{2}$ km/h.   **17** 20.

18 60 mins   1 875 tins per h.   **19** $12\frac{1}{2}$ s.

20 10 min.   **21** 14·4 min   168 letters per min.

22 {a, b, d}   {c, e}.

23 (a) 9:25   (b) 27:125   (c) 3:5.

24 (a) 3:4   (b) 9:16   (c) 27:64.

25   1:10   1:6   3:20   4:5
     1:100   1:36   9:400   16:25
     1:1000   1:216   27:8000   64:125
     2:25   2:7
    16:2500   4:49
     8:15625   8:343.

26 (a) 0·1   0·25   2·5   10 km
     (b) 100   16   0·16   0·01 cm².

27 (a) 1·6 m   (b) 40 000:9.

28 (a) 1:2 500   (b) 1:50   (c) 1:125 000   (d) 1:50
     (e) 1:125 000.

29 $656\frac{1}{4}$ litres;   625:1.

30 4·8 m   9 m²   1·2 m²   1 in 30   4 m³.

Exercise 9

1   0·00   0·50   0·71   0·87   1·00
     1·00   0·87   0·71   0·50   0·00
     0·00   0·58   1·00   1·73 ∞.

2   tan x $\to \infty$ as x $\to 90°$.

3   (a) sin   tan   (b) cos decreases as the angle increases
     (c) sin   cos   (d) yes   (e) 0 to ∞.

4   (a) 45°   (b) sin x   cos x decreases   (c) yes
     (d and e) sin x and cos x are not straight line graphs.

5   0·4766   0·5861   0·7038   0·9046   0·9701   0·9914
     0·5422   0·7234   0·9907   2·121   3·995   7·562.

6   0·8791   0·8102   0·7104   0·4263   0·2427   0·1311
     $\bar{1}$·9441 $\bar{1}$·9086 $\bar{1}$·8515 $\bar{1}$·6297 $\bar{1}$·3853 $\bar{1}$·1176
     yes.

7   a   b   d   e   f.

8   (a) 10°27′   20°08′   38°40′   42°45′   45°49′
     (b) 79°33′   69°52′   51°20′   47°15′   44°11′
     (c) 10°17′   19°00′   32°00′   34°10′   35°39′.

9   (a) 3·94   4·53 cm   (b) 4·93   6·30 cm   (c) 6·25
     11·8 cm   (d) 4·59   10·1 cm   (e) 23°35′
     (f) 48°11′   (g) 56°18′.

10   (a) 7·26 cm   (b) 12·3 cm   (c) 25°40′   (d) 65°22′
     (e) 27·7 cm²   (f) 35·9 cm².

11   (a) 3·36   3·24 cm   (b) 6·25   6·74 cm
     (c) 6·79   5·90 cm   35°44′   (d) 17·1 cm².

12   (a) 9   (b) 9   (c) 40°   80°   120°   (d) all 5 cm
     (e) isosceles   (f) 3·42 cm   (g) 6·43   8·66 cm.

13 $\frac{3}{5}$ $\frac{12}{13}$ $\frac{8}{17}$ $\frac{20}{29}$ $\frac{7}{25}$ $\frac{9}{41}$
     $\frac{4}{5}$ $\frac{5}{13}$ $\frac{15}{17}$ $\frac{21}{29}$ $\frac{24}{25}$ $\frac{40}{41}$
     $\frac{3}{4}$ $\frac{12}{5}$ $\frac{8}{15}$ $\frac{20}{21}$ $\frac{7}{24}$ $\frac{9}{40}$.

14   0 $\frac{1}{2}$ $\frac{1}{\sqrt{2}}$ $\frac{\sqrt{3}}{2}$ 1

      1    $\frac{\sqrt{3}}{2}$ $\frac{1}{\sqrt{2}}$ $\frac{1}{2}$ 0

      0    $\frac{1}{\sqrt{3}}$ 1 $\sqrt{3}$ ∞

15 (a) 20 $\quad$ $10\sqrt{3}$ $\quad$ 10 $\quad$ $5\sqrt{3}$ $\quad$ 15

$\quad$ (b) $4\sqrt{3}$ $\quad$ 6 $\quad$ $2\sqrt{3}$ $\quad$ 3 $\quad$ $3\sqrt{3}$

$\quad$ (c) $\dfrac{20\sqrt{3}}{3}$ $\quad$ 10 $\quad$ $\dfrac{10\sqrt{3}}{3}$ $\quad$ 5 $\quad$ $5\sqrt{3}$

$\quad$ (d) 12 $\quad$ $6\sqrt{3}$ $\quad$ 6 $\quad$ $3\sqrt{3}$ $\quad$ 9

$\quad$ (e) $\dfrac{28\sqrt{3}}{3}$ $\quad$ 14 $\quad$ $\dfrac{14\sqrt{3}}{3}$ $\quad$ 7 $\quad$ $7\sqrt{3}$

16 2·65 $\quad$ 4·24 km.

17 (a) 5·56 $\quad$ 5·75 km $\quad$ (b) 2·18 $\quad$ 8·73 km
$\quad$ (c) 0·872 $\quad$ 9·96 km.

18 140° approx.

19 (a) 149° $\quad$ (b) 034° $\quad$ (c) 310° $\quad$ (d) $243\frac{1}{2}$° approx.

20 150 km. $\qquad$ **21** 217 m $\quad$ 811 m.

22 (a) 110 m $\quad$ (b) 63·3 m $\quad$ (c) 1 070 m.

23 048°22′; the same.

24 (a) 35·8 m $\quad$ (b) 46·7 m. $\qquad$ **25** 4·10 m $\quad$ 2·87 m.

26 43°36′. $\qquad$ **27** 13·3 cm $\quad$ 8·95 cm.

28 18°26′ $\quad$ 14°02′ $\quad$ 11°19′.

29 36·3 m $\quad$ 10·4 m. $\qquad$ **30** 12·4 m.

Exercise 10

1 5 $\quad$ −5 $\quad$ −5 $\quad$ 5; $\quad$ 32 $\quad$ −32 $\quad$ −32 $\quad$ 32;
$\quad$ 3 $\quad$ −3 $\quad$ −3 $\quad$ 3; $\quad$ 35 $\quad$ −35 $\quad$ −35 $\quad$ 35;
$\quad$ 9 $\quad$ 3 $\quad$ 3 $\quad$ 9; $\quad$ 20 $\quad$ 4 $\quad$ 4 $\quad$ 20.

2 9 $\quad$ −27 $\quad$ 6 $\quad$ 10 $\quad$ −30 $\quad$ −6 $\quad$ 4 $\quad$ −1·2 $\quad$ 2·4.

3 −1 $\quad$ 72 $\quad$ 16 $\quad$ −27 $\quad$ $-1\frac{1}{6}$ $\quad$ $-1\frac{1}{2}$ $\quad$ 18 $\quad$ 96 $\quad$ 54.

4 (a) 18 $\quad$ 69 $\quad$ 123 $\quad$ (b) 4 $\quad$ 100 $\quad$ 184 $\quad$ 292.

5 (a) $-\frac{5}{6}$ $\quad$ (b) $\frac{1}{6}$ $\quad$ (c) $-\frac{1}{6}$ $\quad$ (d) 13.

6 $1 - x$; $\quad$ $5 - x$; $\quad$ -4; $\quad$ $x + 7$; $\quad$ $12 - 2x$;
$\quad$ $8x - 12$; $\quad$ $15x - x^2$; $\quad$ $11x^2 - 35x$; $\quad$ $30 - 4x$;
$\quad$ $30x - 72$.

7 (a) $\sqrt{2}$ $\sqrt{3}$ $\sqrt{4}=2$ $\sqrt{5}$ $\sqrt{6}$ $\sqrt{7}$
    (b) $\sqrt{5}$ $\sqrt{9}=3$ $\sqrt{13}$ $\sqrt{17}$ $\sqrt{21}$ $\sqrt{25}=5$
    (c) $\sqrt{5}$ $\sqrt{14}$ $\sqrt{30}$ $\sqrt{55}$ $\sqrt{91}$
    $\sqrt{140}=2\sqrt{35}$.

8 $\sqrt{41}$ $\sqrt{73}$ $\sqrt{40}=2\sqrt{10}$ $\sqrt{130}$ $5\sqrt{5}$
  $4\sqrt{5}$ m.

9   9   3   27   49   7   28   169   13   325 cm².

10   4 $\sqrt{61}$ 5 $2\sqrt{13}$ 6 $3\sqrt{7}$ 9 $3\sqrt{2}$ cm.

11 $\sqrt{2}$ $2\sqrt{2}$ $\sqrt{5}$ $2\sqrt{5}$ $\sqrt{7}$ $3\sqrt{7}$ $7\sqrt{2}$
  $11\sqrt{2}$ cm.

12 $\sqrt{90\,000}=300$; $\sqrt{1{\cdot}69}=1{\cdot}3$; $\sqrt{0{\cdot}0\,169}=0{\cdot}13$;
  $\sqrt{81\,000}=900$.

13   (a) $\sqrt{48}$ (b) $3\sqrt{5}$ (c) $3\sqrt{9}$ (d) $5\sqrt{25}$.

14 $\pm\sqrt{5}$; $\pm6\sqrt{2}$; $\pm2\sqrt{13}$; $\pm2\sqrt{7}$; $\pm5\sqrt{2}$;
  $\pm4\sqrt{2}$.

15 $\frac{1}{2}(-5\pm\sqrt{21})$; $\frac{1}{2}(-5\pm\sqrt{29})$; $\frac{1}{2}(3\pm\sqrt{5})$;
  $\frac{1}{2}(3\pm\sqrt{13})$.

16 $\dfrac{1}{\sqrt{2}}$ $\dfrac{\sqrt{3}}{2}$ $\sqrt{3}$ $\dfrac{1}{\sqrt{2}}$ $\dfrac{1}{\sqrt{3}}$.

17 $\dfrac{16\sqrt{3}}{3}$ cm $\dfrac{32\sqrt{3}}{3}$ cm $\dfrac{128\sqrt{3}}{3}$ cm².

18   8 by $4\sqrt{2}$; $4\sqrt{2}$ by 4;   4 by $2\sqrt{2}$.

19   9 by $3\sqrt{3}$; $3\sqrt{3}$ by 3.

22   (a) wholes   integers   rationals   rationals
     rationals   irrationals
    (b) wholes   irrationals   wholes   rationals
     rationals   wholes
    (c) rationals   rationals   irrationals   irrationals
     irrationals   rationals.

23   (a) rationals   (b) wholes   (c) naturals   (d) integers
    (e) squares   (f) evens   (g) negative integers
    (h) squares.

24 (a) wholes   (b) rationals   (c) rationals
(d) rationals   (e) rationals   (f) integers   (g) wholes
(h) rationals.

25 wholes   rationals   wholes   integers   integers
irrationals   integers   irrationals.

Invent a new number $\sqrt{-15}$ or $\sqrt{-1}\sqrt{15}$.

26 (a) wholes   (b) rationals   (c) rationals
(d) rationals   (e) irrationals   (f) rationals
(g) irrationals and rationals.

Exercise 11

1 Thursday for a b c e   Saturday for d.

2 (a) 1 am   (b) 3 am   (c) 8 pm   (d) 5 pm
(e) 5 am.

3 30°   110°   30°   280°   154°   62°   70°.

4

$+$0	1	2	3		$\times$0	0	0	0
1	2	3	0		0	1	2	3
2	3	0	1		0	2	0	2
3	0	1	2		0	3	2	1

(a) 1 0 2 2 2 2 0 2   (b) 1 3 2 3

(c) no solution 2 no solution 3.

5 No solution 3 1 1 or 3 0 or 2.

6

$+$0	1	2	3	4		$\times$0	0	0	0	0
1	2	3	4	0		0	1	2	3	4
2	3	4	0	1		0	2	4	1	3
3	4	0	1	2		0	3	1	4	2
4	0	1	2	3		0	4	3	2	1

(a) 1 2 2 4   (b) 2 3 3 0 3

(c) 1 2 4 1 4   (d) 3 2 3 2 4 4 1.

7 3 4 4 2 1 2 or 3 1 or 4 4 or 3.

8

x^2	x^3	x^4				
0	0	0	1 or 4	no solutions	2 or 3	2
1	1	1	no solutions	1 or 2 or 3 or 4	0	2.
4	3	1				
4	2	1				
1	4	1				

9

+0	1	2	3	4	5	×0	0	0	0	0	0
1	2	3	4	5	0	0	1	2	3	4	5
2	3	4	5	0	1	0	2	4	0	2	4
3	4	5	0	1	2	0	3	0	3	0	3
4	5	0	1	2	3	0	4	2	0	4	2
5	0	1	2	3	4	0	5	4	3	2	1

(a) 1   4   3   2
(b) 0   1   3   0
(c) 3   4   4   3
(d) 4   4   2   0   4
(e) 2 or 5   1 or 3 or 5   0 or 2 or 4   no solutions.

10   4   3   2 or 5   no solution   2 or 4   1 or 5
2 or 4   no solution   3.

11 (a) E   W   N   S   E   S   S   E
(b) E   S   N   W   W   S.

12 (a) 3   3   3   1   1   1   1   litres
(b) 1   1   0   litres.

13 Table as in question 9
2   5   3   2   4   cubes.

14 (a) 3   (b) 8 tonnes   (c) 12 tonnes   (d) Mod 15.

Exercise 12

1 1   10   11   100   101   110   111   1 000
1 001   1 010   1 011   1 100   1 101   1 110   1 111.

2 13   11   12   21   57   43   438.

3 10 011   10 111   100 011   101 001   111 001
1 000 101   1 001 001   11 000 000.

4 $\frac{1}{2}$ $\frac{3}{8}$ $\frac{5}{8}$ $\frac{7}{8}$ $\frac{5}{16}$ $\frac{21}{32}$.

5 0·1   0·11   0·001   0·01   0·101   0·1001   0·00111

6 1 111·1   10 011·001   10 111·11   11 101·011
100 101·1011.

7 $5\frac{3}{4}$ $6\frac{1}{4}$ $9\frac{5}{8}$ $13\frac{7}{8}$ $10\frac{9}{16}$.

8 (a) 10 010   11 001   10 001   101 000
(b) 1 000 001   1 000 010   10 101 110   10 010 011
(c) 10   10 100   11 100   11 001
(d) 100   100 100   11 011 000   101 010 111.

9 100 100   10 010   11 110 011   11   1 010 001.

10 (a) 1 000 010   110 100   110 011 101   1 000   r 11
110 001
(b) 110 010   11 110   110 010 000   100   1 100 100
(c) 11 000   1 100   1 101 100   11   100 100.

11 (a) 11   (b) 111   (c) −10 or −11   (d) −1 or 100
(e) 100.

12 27   38   51   83   No figure 2 or 3 in base 2.

13 222_8 444_5 111_9 333_4.

14 $1\,011_2$ 33_4 201_3 43_8 101_6.

15 $0 \cdot 22_5$ $0 \cdot 46_8$ $0 \cdot 33_4$.

16 4   8;   no base;   3   5   6   8   9;   4   6;
no base.

17 (a) 11   131   521   (b) $1\,763_8$ $13\,021_5$
$1\,111\,110\,011_2$
(c) If even base then there is an odd 1 at the end
If odd base then it is made up of an odd number of
odd numbers.
(d) No e.g. $1\,111_3 = 40_{10}$.

18 419   1 487   1 713   862   1 571   1 490.

19 1 010   232   21 303   1 321   2.

20 344   243   13 433   1 134   13
344   254   12 223   1 013   1.

21 (a) Ends in 0   00   000   000 000
(b) Ends in 0   00   000   00 000
(c) 1. The odd numbers end in an odd number
2. The odd numbers have an odd number of odd
digits.

22 Base 4   7   7   6   5   10   4   10.

Exercise 13

1 (a) 5.4×10^7 (b) 3.3×10^5 (c) 1.1×10^7
   (d) 2.3×10^8 (e) 9.4×10^2 (f) 1.9×10^5
   (g) 6.6×10^5.

2 (a) 8 420 000   (b) 19 500   (c) 552   (d) 7 700 000
   (e) 59.5   (f) 560 000 000 km².   a and d.

3 (a) 3×10^8 m/s   (b) 33 000 cm/s   (c) 330 m/s.

4 (a) 1 020 000 m³   (b) 3.44×10^5 m³.

5 (a) 385 000 km   (b) 1.49×10^8 km
   (c) 3.85×10^8 m;   385 000 000 m;
     149 000 000 000 m; 1.49×10^{11} m.

6   30 9.11×10^{-28}.   **7** 1.5×10^{-38}.

8 (a) 3×10^8 1.2×10^{-1} 2.5×10^9 3.6×10^7
   (b) 1.2×10^4 7.5×10^{-8} 1.6×10^{11} 9×10^{-4}
   (c) 5.4×10^3 1.5×10^{-14} 3.6×10^{17}
     8.1×10^{-11}
   (d) 3.5×10^{-8} 1.4×10^{-1} 2.5×10^{-7}
     4.9×10^{-9}.

9 (a) 2.4×10^8 (b) 9×10^{14} (c) 5×10^5
   (d) 1×10^{19}.

10 (a) 6.6×10^7 (b) 2.7×10^8 km   (c) 6.4×10^7 cm³
   (d) 3.75×10^{-3} cm; 3.75×10^{-2} mm;
     3.75×10^{-5} m
   (e) 1.5×10^{-6} cm³; 1.5×10^{-3} mm³;
     1.5×10^{-12} m³.
   (f) 6.1×10^5 h.

Exercise 14

2 $15a + 21$; $4a + 8b + 12c$; $15r - 10s + 5t$;
   $2x^2 + 3x$; $12y^2 + 15y$; $10z - 6z^2$;
   $6s^3 - 18s^2 + 24s$.

3 (a) $2x^2 + 5x + 3$ (b) $2x^2 - x - 3$
   (c) $2x^2 - 5x + 3$.

4 $x^2 + 5x + 6$; $y^2 + 3y - 18$; $z^2 - z - 20$;
$3p^2 - 7p - 6$; $2q^2 + 7q - 30$; $5r^2 - 14r - 3$;
$8s^2 - 18s - 35$; $6t^2 + t - 40$.

5 $a^2 + 2ab + b^2$; $a^2 - 2ab + b^2$.

6 $x^2 + 2x + 1$; $y^2 - 4y + 4$; $z^2 + 6z + 9$;
$4p^2 - 4p + 1$; $16s^2 + 24s + 9$; $25t^2 - 70t + 49$;
$a^2 + 2ab + b^2$; $a^2 + 4ab + 4b$;
$4a^2 - 12ab + 9b^2$; $25a^2 - 40ab + 16b^2$;
$16a^2 - 40ab + 25b^2$.

7 $a^2 + b^2 + c^2 + 2ab + 2bc + 2ca$;
$a^2 + 4b^2 + 9c^2 + 4ab + 12bc + 6ca$;
$a^2 + b^2 + c^2 + 2ab - 2bc - 2ca$;
$a^2 + b^2 + c^2 - 2ab - 2bc + 2ca$.

8 (a) $p(q + r)$ 150;   (b) $q(p + q)$ 1 840
(c) $p(q - r)$ 605   298.

9 $x(x + 2)$; $y(y - 5)$; $z(3z + 4)$; $4p(2p + 1)$;
$3q(1 - 3q)$; $4r^2(r + 6)$;
$5s^2(1 - 4s^2) = 5s^2(1 + 2s)(1 - 2s)$; $6t^2(3t + 2)$;
$r(2 + \pi)$; $2\pi(R + r)$; $\pi r(l + r)$;
$2\pi r(r + h)$.

10 $(x + 2)(x + 5)$; $(y + 3)(y + 7)$;
$(z + 4)(z + 9)$; $(p - 6)(p + 2)$;
$(q - 8)(q + 3)$; $(r + 9)(r - 5)$;
$(2s + 5)(s + 1)$; $(3t + 2)(t + 4)$;
$(5n + 3)(n - 2)$; $(2p + 1)(p + 3)$;
$(3q + 2)(2q + 5)$; $(4r + 3)(2r + 3)$;
$(3a + 2)(2a - 3)$; $(4b + 1)(2b - 5)$;
$(3c + 4)(2c - 3)$; $(4d - 1)(3d + 5)$;
$(x + 3)^2$; $(x - 5)^2$; $(x - 4)^2$; $(2x + 1)^2$;
$(3x - 2)^2$; $(n + 3)(n - 3)$; $(q + 11)(q - 11)$;
$(2r + 5)(2r - 5)$; $(3s + 7)(3s - 7)$;
$3(x + 2)(x - 2)$; $2(y + 3)(y - 3)$;
$5(z + 4)(z - 4)$.

11 $(x + 7)(x - 7)$;   7 800   800   30   58   288   212
400.

12 $\pi(R + r)(R - r)$;   176   528   1 210.

13 $(a + b)(c + d)$; $(x + 2)(x^2 + 3)$;
$(p + q)(r + 2s)$; $(r + 2s)(s + 3t)$;
$(p + q)(q - r)$; $(4x - y)(y + 2z)$.

14 (a) $\dfrac{x}{y} = \dfrac{3x}{3y} = \dfrac{8x}{8y} = \dfrac{-5x}{-5y} = \dfrac{x^2}{xy} = \dfrac{xy}{y^2}$

(b) $\dfrac{p}{q} = \dfrac{6p}{6q} = \dfrac{9p}{9q} = \dfrac{p^3}{p^2q} = \dfrac{3pq}{3q^2} = \dfrac{p(p+1)}{q(p+1)} = \dfrac{p(q-4)}{q(q-4)}$

(c) $\dfrac{x-3}{5} = \dfrac{2x-6}{10} = \dfrac{4x-12}{20} = \dfrac{(x-3)(x+2)}{5(x+2)} =$

$\dfrac{x^2-3x}{5x} = \dfrac{x^2-9}{5(x+3)}$

(d) $\dfrac{r+s}{r-s} = \dfrac{3r+3s}{3r-3s} = \dfrac{-r-s}{s-r} = \dfrac{2(r+s)}{2(r-s)} =$

$\dfrac{(r+s)(r-s)}{(r-s)^2}.$

15 $\dfrac{5}{6x} \quad \dfrac{1}{12x} \quad \dfrac{7}{6a} \quad \dfrac{7}{9a} \quad \dfrac{5}{18b} \quad \dfrac{11}{30c} \quad \dfrac{1}{x} \quad \dfrac{28x+45}{72x^2}.$

16 $\dfrac{2(x+1)}{x(x+2)} \quad \dfrac{3}{x(x+3)} \quad \dfrac{2}{x(x-2)} \quad \dfrac{x+3}{2x(x-3)}$

17 (a) $\dfrac{2x+5}{(x+2)(x+3)}$ (b) $\dfrac{1}{(x+2)(x+3)}$

(c) $\dfrac{x+2}{(x+3)(x+4)}$ (d) $\dfrac{x}{(x+1)(2x+1)}$

(e) $\dfrac{1}{2(x-2)}$ (f) $\dfrac{3x+7}{(x+2)(x+3)}$

(g) $\dfrac{x-3}{(5x+1)(3x-1)}.$

18 (a) $\dfrac{2x+1}{x^2-1}$ (b) $\dfrac{4-x}{(x+3)(x-3)}$ (c) $\dfrac{1}{x^2-1}$

(d) $\dfrac{2x}{(2x-1)^2}$ (e) $\dfrac{6}{(x+3)^2(3-x)}.$

19 (a) $\dfrac{x}{2y} \quad \dfrac{2x}{5} \quad \dfrac{3y^2}{2} \quad 2a$

(b) $3 \quad \dfrac{b}{2} \quad -1 \quad -\tfrac{1}{2} \quad \dfrac{1}{x-2} \quad \dfrac{-1}{s+t}$

(c) $\dfrac{q}{p-2} \quad \dfrac{5}{x-3} \quad \dfrac{4}{s+2t}$

(d) $2pr \quad -1\tfrac{1}{2} \quad \dfrac{1}{2(x+3)}.$

Exercise 15

1 6.   **2** 3.   **3** 20.   **4** 4.   **5** 12.

6 3.   **7** 18.   **8** 3.   **9** $3\frac{1}{2}$.

10 3.   **11** -10.   **12** 3.   **13** 2.

14 4.   **15** 3.   **16** $4\frac{1}{2}$.   **17** -4.

18 3.   **19** 7.   **20** 17.   **21** $-1\frac{1}{2}$.

22 (a) $35°$ $60°$ $85°$ (b) $30°$ $50°$ $100°$
(c) $44°$ $44°$ $92°$.

23 (a) $(x + 10)$ m   (b) $(5x + 30)$ m   (c) 15 m
(d) 25 by 15 m.

24 (a) 19 39 159 (b) $(2n + 3)$ $(2n - 1)$
$(6n + 3)$ (c) 45 89 91 93 (d) $6n + 3 = 3(2n + 1)$.

25 280 cm².   **26** 16 km.

27 $x = 10$ $y = 2$.   **28** $p = 6$ $q = 4$.

29 $r = 7$ $s = -2$.   **30** $x = 3$ $y = -3$.

31 $v = 11$ $w = 10$.   **32** $y = 7$ $z = -4$.

33 $x = 2\frac{1}{2}$ $y = 1\frac{1}{2}$.   **34** $a = 6$ $b = 16$.

35 $k = 5$ $l = -5$.   **36** $x = 3$ $y = 4$.

37 $x = 2$ $y = 8$.   **38** $a = 3$ $b = 5$; 29.

39 $a = 5$ $b = -3$; 77.

40 $a = 4$ $b = 60$; 46.   **41** $x = 1$ $y = 5$.

42 $p = 5·8$ $q = 5·4$ **43** £3·20 and 70p.

44 (a) 5p (b) 9p.   **45** (a) 80p (b) $7\frac{1}{2}$p.

46 -1 -2.   **47** 1 4.   **48** 2 4.

49 -3 -4.   **50** -5 3.   **51** -2 9.

52 -1 $1\frac{1}{2}$.   **53** $-1\frac{2}{3}$ 2.

54 $-1\frac{1}{2}$ -4.   **55** -3 $\frac{2}{5}$.   **56** $-\frac{1}{2}$ $\frac{2}{3}$.

57 $-\frac{3}{4}$ 1.   **58** 0 -1.   **59** 0 $-\frac{3}{5}$.

60 0 $1\frac{1}{2}$.   **61** 0 3.   **62** -11 11.

63 -8 8.

64 (a) 4   1; ± 2 ± 1;   (b) $\frac{1}{4}$ 1; $\pm \frac{1}{2}$ ± 1
(c) -2 3;   4   9   (d) -5 2;   4   25
(e) 2   3; $\frac{1}{2}$ $\frac{1}{3}$ (f) $-1\frac{1}{2}$ $\frac{1}{3}$; $-\frac{2}{3}$ 3.

65 (a) $x^2 - 3x - 10 = 0$ (b) $y^2 - y - 12 = 0$
(c) $2z^2 - 13z + 20 = 0$ (d) $3r^2 - 10r + 3 = 0$
(e) $s^2 - 6s = 0$ (f) $4t^2 - 49 = 0$.

66 (a) 78   171   465   5 050   (b) 6   15   20.

67 (a) 2   5   14   54   (b) 8   10   16.
(c) $n^2 - 3n - 80 = 0$ has no whole number solutions.

68 12 cm   60 cm.   **69** 28 cm.

70 16 cm   20 cm.   **71** 9   12   15 cm.

72 (a) 140   192   725 cm²   (b) 20   24 cm;   24
28 cm;   66   70 cm   (c) 6   10 cm;   11   15 cm;
15   19 cm.

Exercise 16

1 (a) $D = 7W$ (b) $S = 3\,600H$
   (c) $R = \dfrac{D}{360}$
(d) $L = 2P$ (e) $B = P(C + T)$
(f) $F = 10G$ or $F = 8G$ (If no thumbs!)
(g) $N = \dfrac{LW}{S^2}$ (h) $t = \dfrac{K}{3}\left[\dfrac{1}{u} + \dfrac{2}{v}\right]$

(i) $L = D(R - 1) + 2S$ (j) $W = \dfrac{M}{T}$
(k) $C = P(2H - 1)$.

2 (a) $V = S^3$ (b) $A = 6S^2$ (c) $L = 12S$
(d) $d = \sqrt{3}S$.

3 (a) $\dfrac{UM}{60}$ $\dfrac{VN}{60}$ km   (b) $\dfrac{UM + VN}{60}$ km
(c) $\dfrac{UM + VN}{(M + N)}$ km/h.

4 (a) BC $\dfrac{N}{P}$ (b) BP BCP (c) $\dfrac{X}{B}$ $\dfrac{S}{PB}$.

5 (a) $\dfrac{4}{3}N\pi r^3$ (b) $\frac{2}{3}H\pi r^3$ (c) $N = \dfrac{3S^3}{4\pi r^3}$.

6 (a) $P = \dfrac{L + D}{D}$ (b) $P = \dfrac{4S}{D}$ (c) $P = \dfrac{2(R + Q)}{D}$

(d) $P = \dfrac{2\pi R}{D}$.

7   8   12   16   44   right angles; $n = \frac{1}{2}(S + 4)$;
   7   14   20   27; $S \geq 2$ and even.

8 (1) $h = \dfrac{2A}{a + b}$ (2) $a = \dfrac{2A}{h} - b$ (3) $b = \dfrac{2A}{h} - a$.

    (1) 72   (2) 16   (3) 24   (4) 16.

9 $\dfrac{C}{2\pi}$; $\dfrac{V}{bh}$ $\dfrac{V}{lh}$ $\dfrac{V}{lb}$; $\dfrac{100I}{rt}$ $\dfrac{100I}{pt}$ $\dfrac{100I}{pr}$;

   $\dfrac{3V}{h}$ $\dfrac{3V}{A}$; $\dfrac{A}{\pi l}$ $\dfrac{A}{\pi r}$; $\dfrac{2S}{a + l}$ $\dfrac{2S}{n} - l$ $\dfrac{2S}{n} - a$;

   $\dfrac{2S}{u + v}$ $\dfrac{2S}{t} - v$ $\dfrac{2S}{t} - u$.

10 $\sqrt{\dfrac{A}{\pi}}$ $\sqrt{\dfrac{A}{4\pi}}$ $\dfrac{E}{C^2}$ $\sqrt{\dfrac{E}{m}}$ $\sqrt[3]{\dfrac{3V}{4\pi}}$ $\dfrac{3V}{\pi r^2}$ $\dfrac{\pi r^2 h}{3}$

   $\dfrac{2s}{t^2}$ $\dfrac{at^2}{2}$.

11 $\dfrac{uv}{u + v}$ $\dfrac{vf}{v - f}$ $\dfrac{uv}{u - v}$ $\dfrac{uf}{u + f}$ $\dfrac{R_1 R_2}{R_1 + R_2}$ $\dfrac{RR_2}{R_2 - R}$

   $\dfrac{RR_1}{R_1 - R}$ $\dfrac{bc}{3b + 2c}$ $\dfrac{2ac}{c - 3a}$ $\dfrac{3ab}{b - 2a}$.

12 (a) $\frac{1}{4}(y - 5)$ (b) $\frac{1}{3}(y + 2)$ (c) $\frac{1}{4}(3y - 2)$
   (d) $\frac{1}{9}(4y + 24)$

   (e) $\pm\sqrt{\dfrac{y - 3}{2}}$ (f) $\pm\sqrt{\dfrac{y + 4}{5}}$ (g) $\pm\sqrt{\dfrac{4y + 5}{3}}$

   (h) $\dfrac{y^2 - 7}{6}$ (i) $\pm\sqrt{\dfrac{y^2 - 4}{3}}$ (j) $\pm\sqrt{\dfrac{y^2 + 3}{2}}$.

13 (a) $\dfrac{2}{3y - 4}$ (b) $\dfrac{7(R + 1)}{R - 1}$ (c) $\dfrac{5s + 4}{2s - 3}$

   (d) $\dfrac{5(1 - p)}{1 + p}$ (e) $\dfrac{5M}{m + 5}$.

14 (a) $\frac{1}{2}(R - 3)$ (b) $2s + 5$ (c) $t - 2$.

Exercise 17

1 (3 a)　(3 b)　(3 c)　(3 d)
(6 a)　(6 b)　(6 c)　(6 d)　The order of the elements
(9 a)　(9 b)　(9 c)　(9 d)　is reversed

2 12;　(3 x)　(3 y)　(3 z);　(1 x)　(2 x)　(3 x)　(4 x);
SR.

3 (m 1)　(m 3)　(m 5)　9;
(u 1)　(u 3)　(u 5)　(a) 3　(b) 3　(c) 1　(d) 0
(g 1)　(g 3)　(g 5);　(1 m)　(5 g)
(1 m)　(1 u)　(1 g)
(3 m)　(3 u)　(3 g)
(5 m)　(5 u)　(5 g)

4 (t u)　(t p)　(a) 2　(b) 3　(c) 2　(d) 4
(o u)　(o p)
(n u)　(n p)
u t　　u o　　u n
p t　　p o　　p n　　　　　63　　511　　15.

5 (2 4)　(2 6)　(2 12)　(2 15)　　No, the order of
(3 4)　(3 6)　(3 12)　(3 15)　　the elements is
(5 4)　(5 6)　(5 12)　(5 15)　　reversed
PQ　　QP　　QP　　PQ　　neither.

6 (a) 11　　(b) 3　　(c) 7　　(d) 2　　(e) 5.

7 No, it is one-many　　P, for example, has two images.

8 (a) one-many　　(b) one-one　　(c) many-many
(d) one-one　　b and d;　$f:x \to 2x - 1$;
$f:x \to x^2$.

9 (a) many-many　　(b) many-one　　(c) one-one
(d) one-many　　b　　c;　$f:x \to \dfrac{1}{x^2}$;　$f:x \to x^3$.

10 1　0　1　0　'attends classes in . . .'
1　0　1　1　(a) G　(b) J　(c) many-many
0　0　1　1　(d) N, for example, has 2 images.

11 (a) many-one　　(b) one-one　　(c) one-one
(d) many-one　　all are functions
(a) 'is classified as' (vowels and consonants)
(b) 'is 3 less than'
(c) 'is the previous letter in the alphabet to'
(d) $f:x \to x^4$.

12 (a) 45   105   105 $-2\frac{1}{4}$ $-2\frac{1}{4}$ (b) many-one
(c) $0 < y < 9$ (d) no.

13 (1a)   18   28 -32 $\frac{1}{2}$ $-4\frac{1}{2}$ (b) one-one
(c) $-12 < y < 8$ (d) yes
(2a) 10 $7\frac{1}{2}$ -15 24   40   (b) one-one
(c) $-\infty < y < 15$ (d) yes

14 (a) 11   23   16   16   (b) f is one-one,
   g is many-one   (c) 4 ± 6
(d) $f:x \rightarrow 3x^2 + 2$; $f:x \rightarrow (3x + 2)^2$.
   yes for f; no for g.

15 (a) 11   32 $\frac{1}{5}$ -3 $\frac{1}{2}$ 5   (b) 21   162 $\frac{2}{25}$
   $\frac{1}{50}$ (c) $\frac{4}{x^2} + 5$; $\frac{1}{x^2} + 5$.

16 Even, square numbers; $f:x \rightarrow 5x$;   whole numbers;
   $0 \leq y \leq 1$;   odd numbers; $0 \leq y \leq \infty$;   integers.

17 (a) $f:x \rightarrow \frac{1}{3}(x - 2)$ (b) $f:x \rightarrow \frac{1}{2}(x + 3)$
(c) inverse is not a function   (d) $f:x \rightarrow 3x - 8$
(e) $f:x \rightarrow \sqrt[3]{x - 4}$.

18 (a) . . . is the brother of . . .
(b) . . . is the sibling of . . .
(c) . . . has less chocolate than . . .
(d) . . . is the daughter of . . .
(e) . . . is half as long as . . .
(f) . . . is $\frac{2}{3}$ as long as . . .
(g) . . . has a factor in common with . . .   Self inverse.

Exercise 18

1 (a) 9   10   17 units of length   (b) 36 units^2
(c) (0 16).

2 (a) (9 0)   (b) $(-9$ 0)   (c) (21 16).

3 (a) 13 units of length   (b) 30   5 units^2
(c) $x = 5$, $y = 2$, $5y = 12x + 10$.

4 (1) 10   (2) 26   (3) -8 or 18.

5 (a) 7   5 units of length   (b) 14 units^2
(c) (14 7);   (8 0);   (0 8)   (d) (7 0).

6 (a) 2   5   3   −2   −1   −1½
  (b) (−1½ 0)   (0 3);   (−⅘ 0)   (0 4);
  (⅓ 0)   (0 −1);   (2 0)   (0 4);   (7 0)   (0 7);
  (4 0)   (0 6)   (c) 2¼   1·6   ⅙   4   24½   12 units².

7 11   −½;   24   −⅖;   11 1;   −4   1;   3   5;
  0   2⅔.

8 (a) (1 5)   (b) 12½ units²   (c) 10¼ units².

9 $y = 2x + 3$; $y = 7 - 2x$.

10 (a) T   (b) T   (c) T   (d) F   (e) T.

11 (a) $x = 3, y = 1$ (b) $x = 2, y = 3$
  (c) $x = -1, y = 4$.

12 (a) £2·80   £2·00   (b) 150 km   140 km
  (c) B by 50p   A by 10p   (d) 133⅓ km.

13 (1) $x = 2, y = 2$ (2) $x = 3, y = -2$
  (3) $x = 6, x = 8, y = 1$
  (4) $x = 6, y = -1, y = -2$
  (5) $y = 24 - 2x, y = 2x - 20$
  (6) $x = 10, x = 12, y = 8 - x$.

14 (a) (0 3)   (4 0)   (8 0)   (0 6)
  (b) 5 units of length   (c) 6 units²   (d) 18 units².

15 (a) (−1 0)   (5 0)   (0 5)   (2 9)
  (b) yes   yes   yes   (c) −1 or 5.

16 (a) −1 or 1½   (b) −1·77 or 2·27   (c) 1·85 or −1·35
  (d) 0 or ½   (e) 1·82 or −0·82   (f) −1·16 or 2·16.

17 (a) 2·89   5·29   10·9   (b) 2·83   3·16   3·87
  (c) −2 or 3   3·56 or −0·56   −3·45 or 1·45.

18 (a) 0 or 4   3·41 or 0·59   1 or 3
  (b) +2   0   −4.

19 4   (a) 8   9   (b) −4.

20 0   30   40   30   0   −50 m.

21 (a) 23 m   (b) 0·8 and 3·2 s   (c) 1·4 s
  (d) downwards and below.

22 (a) 5½   15½   30½ m   (b) 3·5   1·4   1·1 s
  (c) 15 m/s.

23 (a) 150   430   700 m   (b) 16 s   (c) 21·4 m/s
  (d) 30 m/s.

24 (a) 8 km   (b) 1400 h   (c) 5 km/h.

25 (a) 24 km/h   (b) 11·22 h   9 km from Lower Lypp.

26 (a) 1605 h   18 km from Uphill   (b) 3 km approx.

Exercise 19

1 (a) 100   84 $112\frac{1}{2}$ (b) 22   (c) 3x   (d) yes
(e) no.

2 (a) 25   41   629   185   (b) no   (c) yes
(d) $b = \pm c$.

3 (a) $\frac{5}{6}$ $\frac{7}{12}$ 6   (b) yes   (c) no   (d) 6 ($p \neq 0$)
(e) yes.

4 (a) yes   (b) no   (c) yes   (d) no   (e) no   (f) yes.

5 (a) yes   (b) no   (c) no   (d) yes   (e) yes   (f) yes.

6 (a) 0   (b) 0   (c) 1   (d) $\begin{pmatrix} 0 & 0 \\ 0 & 0 \end{pmatrix}$ (e) $\begin{pmatrix} 1 & 0 \\ 0 & 1 \end{pmatrix}$

7 (a) −6   (b) $\frac{1}{12}$ (c) $\frac{8}{5}$ (d) $\begin{pmatrix} -4 \\ 2 \end{pmatrix}$

(e) $\begin{pmatrix} -2 & -3 \\ -1 & -4 \end{pmatrix}$ (f) $\begin{pmatrix} 2 & -1 \\ -5 & 3 \end{pmatrix}$ (g) letter M.

9 (a) no   (b) no   (c) no.

10 (a) yes   (b) yes.

11 Closed, associative, an identity element and inverse elements.   No.

12 1   3   5   yes   yes   yes   no inverse for 3.
　3   3   3
　5   3   1

13 1   3   5   7
　3   1   7   5
　5   7   1   3
　7   5   3   1

14 (a) 7   (b) yes   (c) 7   (d) 1   3   5   or   7.

15
```
1  3  5  7  9      yes    yes    no inverse for 5.
3  9  5  1  7
5  5  5  5  5
7  1  5  9  3
9  7  5  3  1
```

16
```
4  8  2  6
8  6  4  2
2  4  6  8
6  2  8  4
```
(a) yes   (b) yes   (c) 6   (d) 4   2
(e) 4   6   (f) yes   (g) yes.

17
```
 0  4  8 12
 4  0  4  8
 8  4  0  4
12  8  4  0
```
(a) yes   (b) no   (c) yes   (d) yes   (e) no.

18 (a)
```
A  S  L  R
S  A  R  L
L  R  S  A
R  L  A  S
```
(b) S   (c) L   (d) A and S
(e) yes   (f) yes

19 (a)
$$
\begin{array}{cccc}
S_1 & S_3 & S_3 & R \\
S_3 & S_2 & S_3 & R \\
S_3 & S_3 & S_3 & R \\
S_1 & S_2 & S_3 & R
\end{array}
$$
(b) yes   (c) no
(d) not associative; not all elements have inverses.

20
```
2  N  5  3  4  1
N  1  4  5  3  2
4  5  N  1  2  3
5  3  2  N  1  4
3  4  1  2  N  5
1  2  3  4  5  N
```

22 The group of Q21 is a subgroup of Q22.

25 (a)
$$
\begin{array}{cccc}
N & B & Z_2 & Z_1 \\
B & N & Z_1 & Z_2 \\
Z_2 & Z_1 & N & B \\
Z_1 & Z_2 & B & N
\end{array}
$$
(b) yes   (c) N
(d) each is self inverse.
(e) yes

26 a   d   b not associative   c not closed
e no inverse for 0.

27
$$
\begin{array}{cccccc}
0 & 1 & 2 & R_2 & R_3 & R_1 \\
1 & 2 & 0 & R_3 & R_1 & R_2 \\
2 & 0 & 1 & R_1 & R_2 & R_3
\end{array}
$$
yes   yes

28
0	1	2	3	R_2	R_3	R_4	R_1	yes	yes.
1	2	3	0	R_3	R_4	R_1	R_2		
2	3	0	1	R_4	R_1	R_2	R_3		
3	0	1	2	R_1	R_2	R_3	R_4		

29
1	2	3	4	yes	no.
2	4	1	3		
3	1	4	2		
4	3	2	1		

30
S	R	L	A	yes	$S \rightarrow 1$
R	A	S	L		$R \rightarrow 2$
L	S	A	R		$L \rightarrow 3$
A	L	R	S		$A \rightarrow 4$

31 (a) L T   (b) yes   (c) yes.
      T L

32 a c and e ; d and f.

Exercise 20

1 (a) yes   (b) no   pineapple ring is different
(c) no, key rings are usually one length of wire.

2 (a) mint-with-hole   (b) British doughnut
(c) tumbler.

3 {diamond, pearl}   {bracelet, bead, ring}

4 (a) Hoop is a torus   string is a sphere
(b) no difference.
(c) pullover is a sheet with 3 holes, cardigan is a sheet with 2 holes.

5 {I C}   {M W}   {T Y J}   {V}

6 (a) Almost any fruit   (b) None seem to grow this way.

7 {a f} {b d}.   **8** {a c e}   {b d}

9 b d f.   **10** a c d e.

11 Yes   outside x   inside y and z.

12 Yes   inside x   outside y and z.

13 (i) 2 2 0 1 3   (ii) d   (iii) b c d.

14 (b) 2   (c) When two houses are linked to T a closed
curve is formed with T inside and the
third house outside.

15 (a) 2 2 2   (b) 4 2 4   (c) 2 1 1
(d) 5 5 8 $N + R - A = 2$ in each case.

16 (a) 12 7 17   (b) 3 5 6   (c) 5 5 8
(d) 6 5 9 $N + R - A = 2$ in each case.

17 (15) a b c   (16) b c d.

18 (i) a b   (ii) c 1 ; d 1.

19 3 arcs meet at each node   4 4 6   6 5 9
3 3·6   12 12 12   18 18 18
$3N = 2A = aR.$

20 8 6 12   4   24 24 24 $3N = 2A = aR.$

22 4 6 8 2   5 7 10 2 $\frac{8}{3}$ $\frac{20}{7}$ 16 16 16
20 20 20.

23 6 8 12 2 3   24 24 24.

24 $a = 4 - \dfrac{8}{R}.$

25 (a) Yes   (b) Yes   D or E
(c) A to D or D to E.

26 (b) Yes   (c) No   link to barge No. 2.

27 (a) 8 6 12   (b) 5 5 8   (c) 4 4 6
(d) 7 7 12   (e) 12 8 18   All 2.

28 (a) $8 + 6 - 12 = 2$ (b) $10 + 7 - 15 = 2$
(c) $16 + 10 - 24 = 2.$

29 (a) 2   (b) 3   (c) 3   (d) 3   (e) 4.

30 (a) 2   (b) 3   (c) 4   (d) 3   (e) 4.

31 The triangular prism.

32 Yes a and c   (a) 3   (b) 3   (c) 3.

33 Yes a and c   (a) 2   (b) 2   (c) 2.

34 Both 4.

35 4 Yes.   **36** (b) 2   (c) one piece.

37 (a) 4   (b) 3   (c) 2   (d) 4.

38 a and e   The two portions have different cyclical orders

eg   1   and   1

$\quad$ 3 2 $\qquad$ 2 3

41 (a) 3   4   (b) (A_4 A_1 A_5 A_2 A_3)   and (A_2 A_3 A_5 A_1 A_4)   (c) A_4 to A_2.

42 (a) Yes   no   yes   no   yes   no
(b) 3   (c) 5(A   C   D   E   B)
(d) 1   make the road opposite A two way
(e) Change to A

$\qquad\qquad$ F.

Exercise 21

1 5·824 ha   1 770 m.

2 45 632 m²   680 m   £5·70   £6·84.

3 18 m²   152 m²   54 m².

4 (a) 6   20   26   cm²   (b) 12   28·7 cm.

5 (a) 68 cm²   (b) 28 cm   (c) 44·4 cm.

6 (a) 144 cm²   (b) 12 cm.

7 (a) 384 m²   (b) 80 m.

8 70 cm²   No, more information needed.

9 (a) 9   (b) $11\frac{1}{2}$ (c) 9   (d) $3\frac{1}{2}$ units²   33 units²

10 $12\frac{1}{2}$ units².   **11** $20\frac{1}{2}$ units².

12 (a) 41·6 cm²   (b) 110 cm²   (c) 618 cm²
(d) 407 cm².

13 (a) 10   (b) 8   (c) 16   (d) 27 units².

14 8·8 km   13·9 km².   **15** 174 km/h.

16 51 100   136 000 approx.

17 (a) 165 m   (b) 176 m   (c) 198 m.

18 (a) 400 m   (b) 422 m   (c) 10 150 m²
(d) 1 440 m² approx.

Exercise 22

1 (a) 145° (b) 163° (c) 145° (d) 135° (e) 162°
(f) 115°.

2 065°.

3 6 n miles.

4 60°   90°   150°   120°   30°.

5 (a) 30° (b) 105°   67½°   127½°   130°.

6 (a) q = 70°   s = 36°   t = 144°
(b) r = 81°   p = 48°   t = 132°
(c) s = 46°   q = 103°   p = 46°
(d) r = 78°   s = 47°   t = 133°
(e) q = 70°   p = 54°   s = 54°

7 (a) 65°   65°   50° (b) 64°   58° (c) 68°
(d) 36.

8 (a) 44°   48° (b) 38°   66° (c) 41°   57° (e) 30.

9 (a) 76°   76°   28° (b) 64°   52°   12°
(c) 69°   27°   111° (d) 36.

10 (a) 18 (b) 24 (c) 45 (d) 9.

11 (a) 140° (b) 150° (c) 156° (d) 160° (e) 162°.

12 (a) 20°   40° (b) 15°   30° (c) 12°   24°
(d) 10°   20° (e) 9°   18°   153°.

13 135°.   **14** 100°.   **15** 30°   60°   150°   120°

16 30°   75°   75°.

17 (a) 30°   45°   105° (b) 90°   90°   90°   90°
(c) 75°   75°   105°   105°.

18 (a) congruent (b) neither (c) congruent
(d) neither (e) similar (f) neither (g) congruent
(h) similar.

19 (a) $\begin{cases} PQ'S \\ PQS \end{cases}$ $\begin{cases} Q'RS \\ QRS \end{cases}$ $\begin{cases} PQ'R \\ PQR \end{cases}$

(b) $\begin{cases} PQS' \\ PQS \end{cases}$ $\begin{cases} QS'R' \\ QSR \end{cases}$ $\begin{cases} PQR' \\ PQR \end{cases}$

(c) $\begin{cases} P'QS' \\ PQS \end{cases}$ $\begin{cases} QRS' \\ QRS \end{cases}$ $\begin{cases} P'QR. \\ PQR \end{cases}$

20 (a) 15 cm   (b) 13 cm   (c) 5:9   (d) 5:14
(e) 39:25.

21 (a) 15   25 cm   (b) 210 cm²   (c) 12 cm
(d) $7\frac{1}{7}$ cm.

23 75 m.   **24** 1·5 m   0·6 m   5·6 m.

25 6   16   24 cm.

26 (a), (b) Triangles PQR   RQS   PRS   (c) 4 $6\frac{2}{3}$ $5\frac{1}{3}$ cm.

27 (a) 10   20 cm   (c) 5:4   5:4   (d) 25:16   4:5
4:9   (e) 20:29.

29 Rhombi

30 (a) parallelograms   (b) rectangles   (c) kites
(d) rhombi   (e) rectangles   (f) rhombi.

33 (a) 6   16 cm   (b) 32·4 $13\frac{1}{3}$ cm   (c) 4   12 cm
(d) 2:5   2:3   4:25   (e) 9:25   16:25.

34 (a) 10 cm   (b) $5\frac{5}{11}$ cm   (c) 16 cm   (d) 15 cm
(e) 15 cm.

Exercise 23

3 5·1 cm.

4 4·5 km   Start the fourth boat at any point on the
circumcircle.

6 90°   45°   45°   The angles are (180° – the given
angles). One axis of symmetry.

7 12·7 cm.   **8** 6   8   10   15.   **10** 1:4:9.

13 A parallelogram.   **14** A parallelogram.

15 (a) 7·9 n miles   294°   (b) 4·2 n miles
34 min   (54 min altogether)   (c) 9·2 n miles   070°.

16 51 m.

17 The locus is a circle radius 4 cm, centre 2 cm from P in OP
produced.

18 (a) a circle   (b) the perpendicular bisector of the line
(c) one point only – the circumcentre
(d) yes – if the points lie on a circle.

Exercise 24

1 180   210   216   400 cm³.

2 40   64   96   64 cm³.

3 (a) 8 400 cm³   (b) 2 000 cm².

4 (a) 6 cm²   (b) 22 cm²   (c) 63 cm².

5 (a) 208 cm²   (b) Join BC to HE
(e) The solid in Fig. 22.3b   (f) 256   356 cm².

6 (a) 528 cm³   (b) 2·64 cm³   (c) 428 000 cm³.

7 SPICK.

8 1 540 cm³   (a) 770 cm³   (b) 385 cm³   (c) $192\frac{1}{2}$ cm³.

9 (a) 1 570 m³   (b) 55 400 m³   (c) 94·2 mm³.

10 (a) 804 cm³   (b) 402 cm³   (c) 101 cm³.

11 13 300 m³.   **12** 111 000 m³.   **13** 88 cm³.

14 144   72   36   36   36 cm³.

15 (a) 120   120   240   240 cm³
(b) 120 cm³ 5 other – 6 in all.

16 (a) 1 020 cm³   (b) 1 360 cm³   (c) 814 cm³.

17 235 cm³.

18 (a) 54 cm   (b) 104 000 cm³ or $\frac{1}{10}$ m³.

19 (a) 38·8 cm³   (b) 11 500 cm³   (c) 4 850 cm³
(d) nil.

20 (a) $1 \cdot 10 \times 10^{12}$ km³   (b) $8 \cdot 56 \times 10^7$ km².

21 16 cm.   **22** 103 cm³.   **23** The hemisphere.

24 48.   **25** 24·3 m³.

Exercise 25

4 90°   acute   90°   acute   acute   90°.

5 AH   HC   EG   AG.   **6** G.

7 equal.   **8** Yes – where diagonals intersect.

9   BCGF   ABFE   HFBD   ABCD   ABGH.

10   21°48′   36°52′   19°26′.

11   17   20   18·0 cm.

12   56°19′   26°20′.

13   (a) 29 cm   (b) 20·5 cm   (c) 18·0 cm.

14   Isos, right angled   scalene right angled
isos, not right angled   scalene, right angled.

15   (a) Isos, right angled   scalene, right angled
scalene, not right angled   scalene, right angled
(b) scalene, right angled   scalene, right angled
scalene, not right angled   scalene, right angled.

16   25°38′   11°19′   10°28′.

17   (a) 50°12′   (b) 58°   (c) 63°26′.

18   (a) 13   15   15·8 cm   (b) 22°37′   29°03′
18°26′   (c) 81°10′.

19   (a) 5   12·65   13 cm   (b) 36°52′   13°21′.
(c) 45°14′.

20 $\sqrt{p^2 + q^2}$ $\sqrt{q^2 + r^2}$ $\sqrt{p^2 + r^2}$ $\sqrt{p^2 + q^2 + r^2}$.

21   (a) 3 cm   (b) 7 cm   (c) 9 cm   (d) 11 cm
(e) 12 cm.

23   Isos, not right angled   isos, not right angled
scalene, not right angled   scalene, right angled.

24   VDC   VAD   ABC   VAC.

25   10   13   12·4 cm.

26   36°52′   67°23′.   **27**   75°58′   67°23′.

28   As question 23
VAB   VBC   VCD and VDA are congruent,
ABC   VAC;   11·3   9   8·06 cm;   45°   51°04′;
60°15′   51°04′.

29   64π   51π   19π cm².

30   75π   50π   25π cm².

31   4·8 cm.

32   4·62   6·21 cm.

Contents of Book 2